M336
Mathematics and Computing: a third-level course

GROUPS & GEOMETRY

UNIT IB I
TILINGS

Prepared for the course team by
Fred Holroyd

The Open University

This text forms part of an Open University third-level course.
The main printed materials for this course are as follows.

Block 1
Unit IB1	Tilings
Unit IB2	Groups: properties and examples
Unit IB3	Frieze patterns
Unit IB4	Groups: axioms and their consequences

Block 2
Unit GR1	Properties of the integers
Unit GR2	Abelian and cyclic groups
Unit GE1	Counting with groups
Unit GE2	Periodic and transitive tilings

Block 3
Unit GR3	Decomposition of Abelian groups
Unit GR4	Finite groups 1
Unit GE3	Two-dimensional lattices
Unit GE4	Wallpaper patterns

Block 4
Unit GR5	Sylow's theorems
Unit GR6	Finite groups 2
Unit GE5	Groups and solids in three dimensions
Unit GE6	Three-dimensional lattices and polyhedra

The course was produced by the following team:

Andrew Adamyk (BBC Producer)
David Asche (Author, Software and Video)
Jenny Chalmers (Publishing Editor)
Bob Coates (Author)
Sarah Crompton (Graphic Designer)
David Crowe (Author and Video)
Margaret Crowe (Course Manager)
Alison George (Graphic Artist)
Derek Goldrei (Groups Exercises and Assessment)
Fred Holroyd (Chair, Author, Video and Academic Editor)
Jack Koumi (BBC Producer)
Tim Lister (Geometry Exercises and Assessment)
Roger Lowry (Publishing Editor)
Bob Margolis (Author)
Roy Nelson (Author and Video)
Joe Rooney (Author and Video)
Peter Strain-Clark (Author and Video)
Pip Surgey (BBC Producer)

With valuable assistance from:

Maths Faculty Course Materials Production Unit
Christine Bestavachvili (Video Presenter)
Ian Brodie (Reader)
Andrew Brown (Reader)
Judith Daniels (Video Presenter)
Kathleen Gilmartin (Video Presenter)
Liz Scott (Reader)
Heidi Wilson (Reader)
Robin Wilson (Reader)

The external assessor was:

Norman Biggs (Professor of Mathematics, LSE)

The Open University, Walton Hall, Milton Keynes, MK7 6AA.

First published 1994. Reprinted 2001, 2005, 2009.

Copyright © 1994 The Open University

All rights reserved. No part of this publication may be reproduced, stored in a retrieval system or transmitted in any form or by any means, without written permission from the publisher or a licence from the Copyright Licensing Agency Limited. Details of such licences (for reprographic reproduction) may be obtained from the Copyright Licensing Agency Ltd of 90 Tottenham Court Road, London, W1P 9HE.

Edited, designed and typeset by the Open University using the Open University TeX System.

Printed in Malta by Gutenberg Press Limited.

ISBN 0 7492 2159 3

This text forms part of an Open University Third Level Course. If you would like a copy of *Studying with The Open University*, please write to the Central Enquiry Service, PO Box 200, The Open University, Walton Hall, Milton Keynes, MK7 6YZ. If you have not already enrolled on the Course and would like to buy this or other Open University material, please write to Open University Educational Enterprises Ltd, 12 Cofferidge Close, Stony Stratford, Milton Keynes, MK11 1BY, United Kingdom.

1.3

CONTENTS

Study guide	4
Introduction	5
1 Tilings	**7**
1.1 The definition of a tiling	7
1.2 The parts of a tiling	11
1.3 Adjacency and incidence	13
2 Tilings using polygons	**16**
2.1 Polygonal tilings	16
2.2 The Archimedean tilings	18
3 Affine transformations and isometries	**22**
3.1 Coordinate systems and linear transformations	23
3.2 Affine transformations	24
3.3 Isometries	26
3.4 A characterization of isometries	27
4 Tilings using congruent polygons	**32**
4.1 Monohedral tilings	32
4.2 The Laves tilings	38
5 Working with isometries	**42**
5.1 A geometric classification of isometries	42
5.2 The Isometry Toolkit	49
Appendix: proof of the fundamental theorem of affine geometry	54
Solutions to the exercises	56
Objectives	66
Index	67

STUDY GUIDE

Make sure that you have read the *Course Guide* before you begin to study this first unit of the course.

This first unit contains rather a large number of definitions, though you should find that they are all fairly straightforward.

It is quite possible that you will have met the material in Section 3 in your previous mathematical studies, though Sections 1, 2, 4 and 5 are probably largely new to you.

You will need your *Geometry Envelope* when you study Sections 2, 4 and 5; this is a large envelope containing a number of A4 cards and overlays. Please *keep it safe* — it is a vital study component that you will need in conjunction with several of the units in the course.

You will need the techniques of Section 5 when you work through *Unit IB3* and *Units GE2–GE4*. If you are short of time, you could postpone your study of this section until you need these techniques. The techniques are summarized in the Isometry Toolkit, which is one of the A4 cards in the *Geometry Envelope*.

The video programme associated with this unit is VC1A *Living with Patterns*, the first programme on Video-cassette 1. You could conveniently view it at any stage in your study of this unit.

There is no audio programme associated with this unit.

INTRODUCTION

This course is concerned with the analysis and classification of certain types of pattern. In the main, the Geometry units of the course (*Units IB1, IB3* and *GE1–GE6*) concern physical patterns, while the Groups units (*Units IB2, IB4* and *GR1–GR6*) are about the more abstract patterns that mathematicians call *groups*. As you will see, there is a strong interplay between the two streams; in particular, the main tool which we shall use in analysing physical patterns is the abstract concept of a *symmetry group*.

You should have met groups and symmetry groups in your previous studies. The concepts are revised in *Unit IB2*.

All human societies of which we have records have used physical patterns in their artefacts. The room in which you are probably sitting almost certainly has many examples of physical patterns. Obvious examples are provided by wallpaper and carpets; less obvious ones, perhaps, by floorboards and brickwork. The material of the clothes you are wearing will have a pattern in the weave, and there may well be an independent, more visible pattern of colours superimposed on the weave pattern.

Figure 0.1 Examples of physical patterns.

The particular types of pattern which we have chosen in introducing the course are *tilings*, partly because of the beautiful examples that are available and partly because it is reasonably easy to define mathematically what we mean by a tiling.

A mathematical definition is given in Subsection 1.1.

The Islamic civilization developed tiling into an art form of amazing intricacy and beauty, used in buildings from the Alhambra in Granada, Spain to the Taj Mahal in India and on to Samarkand and Mongolia.

Johannes Kepler (1571–1630) started to develop the mathematical theory of tiling patterns, and this development was taken up by crystallographers such as the Russian Evgraf Fedorov (1853–1919) in the nineteenth century and the German Fritz Laves early in the twentieth century.

A tiling pattern can be considered as a two-dimensional analogue of a three-dimensional crystal structure.

Recent years have seen a great increase in our theoretical understanding of tiling patterns, culminating in the classifications given by Branko Grünbaum and Geoffrey Shephard in the 1970s and 1980s, and in several quite new types of tiling pattern, most notably Roger Penrose's tilings using two shapes with which it is possible to tile the whole plane in infinitely many distinct ways but impossible to produce a repeating pattern.

In this unit we look at some basic properties of tilings.

In Section 1 we introduce the mathematical definitions of a *tiling* and the *parts* of a tiling, and we consider how the parts are related.

The analysis of *non-periodic tilings* (i.e. those in which there is no repeating pattern) is an interesting area of study, but is beyond the scope of this course, which restricts itself to repeating patterns. If you wish to read more about non-periodic tilings, we would recommend *Tilings and Patterns*, by B. Grünbaum and G.C. Shephard (see the Bibliography in the *Course Guide*). This excellent book also covers repeating patterns, and many of the diagrams in the Geometry units and on the cards in the *Geometry Envelope* are based on diagrams in this book.

In Section 2 we restrict our attention to particular tilings in which each tile is a polygon, and we introduce an important class of polygonal tilings, known as the *Archimedean* tilings.

In Section 3 we begin to set up the algebraic techniques that you need in order to study the symmetries of tilings and other patterns. We consider *affine transformations* of the plane (which can be described by functions of the form $\mathbf{x} \mapsto \mathbf{Ax} + \mathbf{p}$) and, in particular, *isometries*, namely distance-preserving affine transformations.

In Section 4 we return to polygonal tilings, and consider those in which all the tiles are congruent.

In Section 5 we refer back to the topic of Section 3, derive a complete geometric classification of isometries of the plane, and describe the Isometry Toolkit, an A4 card contained in the *Geometry Envelope* and on which are given all the equations you should need in order to manipulate isometries.

1 TILINGS

1.1 The definition of a tiling

The idea of a tiling pattern is intuitively quite a natural one; it is a pattern formed by tiles covering an area such as a wall or a floor. However, this description is too vague to be of much use to mathematicians. We need to develop a more formal definition, and a good way of introducing it is to ask you to see whether *your* intuitive idea of a tiling pattern corresponds to the definition which we shall finally use.

Which of the pictures in Figure 1.1 would you regard as depicting or defining a tiling pattern?

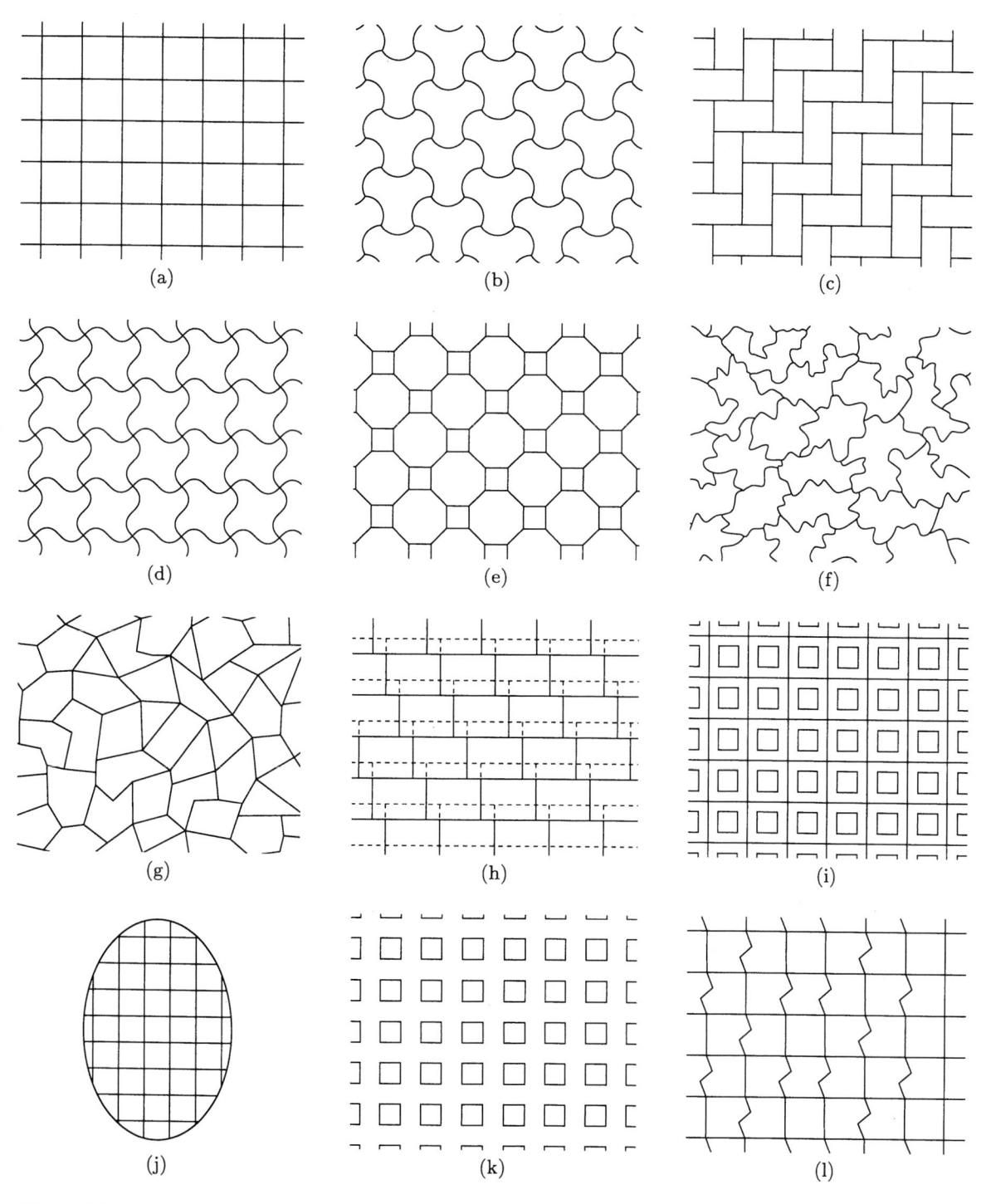

Figure 1.1

Not everyone will come up with the same answers, as there is a certain amount of ambiguity in the idea of a 'tiling'. Do the individual tiles all have to be the same shape? Can they overlap? Do they have to be straight-sided? Does the pattern made by the tiles have to be 'regular'? If so, then in what sense? Does the tiling have to cover the whole plane?

In mathematics, definitions are acceptable provided that they make sense and do not leave any 'borderline cases' that one is not sure how to classify. This leaves plenty of elbow-room; definitions of a 'tiling' could be constructed that would give virtually any combination of positive and negative answers to the above questions.

If you assumed that the individual tiles all have to be the same shape, then you should certainly have rejected (e), (f), (g), (j) and (l). As for (i), what exactly *is* a single tile in this case? If you assumed that tiles must not overlap, then you should have rejected (h). If tiles have to be straight-sided, then (b), (d) and (f) must be rejected, as must (j) because of the curved boundary. If the pattern must have a recognizable regularity, then (f) and (g) must be rejected. If it must cover the whole plane, then (j) must be rejected as it covers only a small elliptical area, while (k) has to go as these 'stepping stones' have gaps between them.

You may well have been puzzled by (i). Does this represent a tiling by large squares each with a small square painted on? Does it represent a tiling by cloister-shaped tiles, each with a small square hole in the middle? Or does it represent a tiling in which some of the tiles are small squares and others are cloister-shaped tiles? Tiles with patterns painted on are presumably acceptable; but cloister-shaped tiles, or any tiles with holes in, are surely not? You may well have decided that (i) is unacceptable, even if you were unable to state why in mathematical terms.

A precise definition of an acceptable shape for a tile would require a detour through the study of topology, which would be rather peripheral to this course, so we shall treat this concept intuitively. Look at the shapes in Figure 1.2.

Shapes (a), (c), (d) and (g) will be regarded as acceptable; in each of these cases the shape has a well-defined inside and a well-defined outside (unlike (h), which has no inside). The fact that the boundaries of (a), (d) and (g) are curved while that of (c) consists of straight lines does not concern us. However, we do require that an acceptable shape should have no holes — even if the hole touches the outside boundary — so (b) and (e) are unacceptable. We also reject tiles with infinitely thin 'spikes' (such as (f)) as unacceptable — such a spike would not hold together physically. Similarly, (i) and (j) would not hold together — in each case it would be possible to separate the tile into two pieces by deleting a single point. Finally, the tile must be of finite extent; infinitely long strips such as (k) are definitely not fair!

Another way of putting this is to say that an acceptable shape for a tile is any flat shape into which a very flexible rubber disc can be stretched without tearing or sticking. Such a shape has a mathematical name: it is a **topological disc**. Thus, the only shapes in Figure 1.2 that are topological discs are (a), (c), (d) and (g). Shape (b) could only be produced by tearing a hole in the disc. Shape (e) could be produced either by tearing a hole or by sticking two points on the boundary of the disc together. Shapes (f), (h), (i) and (j) all have single points whose deletion would separate the material into two parts; as this is not true of an unstretched disc, it cannot be true of a stretched one. Finally, no amount of stretching can actually reach infinity, so (k) is not a topological disc.

Topology is the study of those properties of shapes that do not change when continuous and reversible transformations are applied to them.

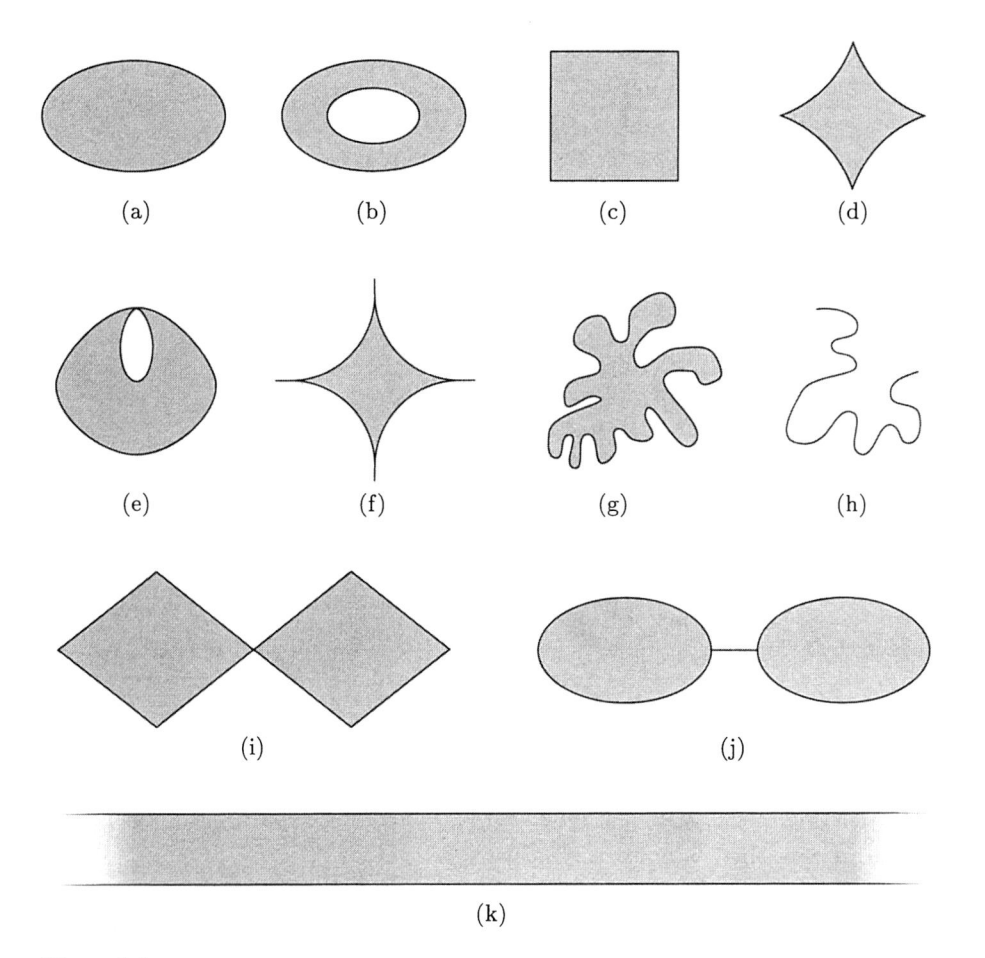

Figure 1.2

Exercise 1.1

Which of the shapes in Figure 1.3 are topological discs?

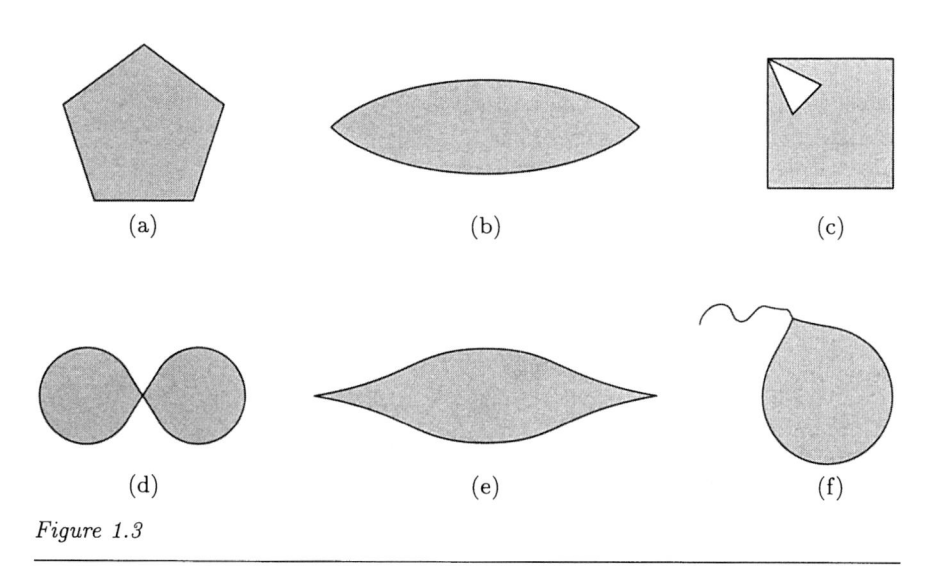

Figure 1.3

Exercise 1.2

Which of the following subsets of the plane are topological discs?

(a) The x-axis.

(b) The set of points whose distance from a fixed point is at most 2 cm.

(c) The set of points whose distance from a fixed point is exactly 2.5 cm.

(d) The set of points lying within, or on the boundary of, an equilateral triangle.

A subtle point arises here. Suppose that option (d) of Exercise 1.2 had specified 'The set of points lying within, but *not* on the boundary of, an equilateral triangle'? This would have described a shape with just as good a case to be called a topological disc as the shape we actually described. So, do we include the *points on the boundary* of a topologically disc-like shape as being a part of the disc, or not? If this were a topology course we would accept *both* possibilities, calling a shape *with* its boundary points a *closed* topological disc and a shape *without* its boundary points an *open* topological disc. But, as this is *not* a topology course, we shall simply choose one option and stick to it. The option we choose is to *include* the boundary points as part of a topological disc.

We must now choose a once-and-for-all definition of *tiling*. It needs to be quite a broad definition (we can always specialize to particular types of tiling as required); but we do not wish it to be so broad as to include things that it would be difficult to say anything about, or that do not fit in with everyday ideas of what tilings are. And we must bear in mind our earlier observation: our definition must not leave any borderline cases that we do not know whether to classify as tilings or not.

The practical function of tiles is to cover a surface. There are plenty of real-life examples of tilings: mosaics, street cobbles, pavings (whether regularly patterned or crazy pavings), roof, floor, wall or even ceiling tilings in buildings, and so on. In the case of roof tilings, the individual tiles are designed to overlap, but the visual effect they produce is of a portion of the plane divided up into *non-overlapping* areas (see Figure 1.4). Sometimes (on floors, for example), tiles of more than one type make up the pattern (see Figure 1.1(e)).

Figure 1.4

In each of these examples *the pattern could in theory be extended to cover the whole plane*. In practice, any tiling is finite, of course, but if you do not know in advance how large the area is that must be tiled, then you have to know that the tiles *could* tile an arbitrarily large area if necessary. Thus, demanding that a tiling should cover the *whole* plane is not quite as abstract as might seem at first sight; this condition does have practical value, as well as being mathematically convenient.

In any case, among mathematicians who study tilings, there is a general agreement to define tilings in such a way that the tiles do cover the whole plane, without overlapping or leaving gaps. Moreover, each individual tile must be a topological disc in the sense we have just defined. These are the only restrictions on the general concept of a tiling, although there are many special types of tiling, some of which you will meet in this unit.

We thus arrive at our formal definition.

Definition 1.1 Tiling

A **tiling** is a covering of the whole plane with non-overlapping **tiles**, each of which is a topological disc.

Non-overlapping means that the tiles meet each other along boundary lines only, so that every point on the plane is either within one single tile or on the boundary of more than one tile. The individual tiles can be of different shapes and can have curved sides; moreover, they need not be arranged in any recognizable pattern, although in most of our examples there *will* be a straightforward pattern.

Exercise 1.3

Which of the pictures in Figure 1.1 represent tilings?

Interpret picture (i) as a pattern of squares and cloister-shapes, where each cloister-shaped tile has a square tile inserted in its central hole.

As you may have realized in solving Exercise 1.3, the pictorial representation of tilings presents something of a problem. A tiling covers the whole plane, whereas any picture covers only a finite area. Clearly, therefore, it can directly represent only a tiny portion of the whole tiling. How can we infer from that portion what the infinite remaining part of the tiling looks like?

Strictly speaking, we *cannot* infer just from a finite picture what the whole of a tiling is like. The picture must be interpreted in its context. Figures 1.1(a)–1.1(e), for example, all suggest patterns which could be repeated indefinitely to cover the whole plane. It would be quite possible, however, for the patterns to break down somewhere in the plane beyond the small portions shown in these pictures; but it is the possible regularity rather than the possible irregularity that is stressed by the pictures, so the natural assumption is that they are pictures of typical portions of tilings with regular patterns. Figures 1.1(f) and 1.1(g), on the other hand, clearly suggest tiles of irregular shapes, arranged according to no particular pattern, although it would be perfectly possible to embed them as part of a large repeating pattern. Whether they are parts of large repeating patterns or not, however, the portions shown do not allow us to infer the patterns of the whole tilings of which they may be parts.

Thus, a *picture* of a tiling is not a mathematically rigorous way of *defining* a tiling. Nevertheless, in practice it is often the most convenient way of conveying which particular tiling we are talking about, and it is the way which we shall normally use. We shall always draw a 'typical' portion of the tiling. That is to say, in those cases (the vast majority we shall consider) where the tilings do have a repeating pattern, the picture will show enough repetitions of the pattern to make it clear how the repetition works.

We shall not, therefore, represent tilings by pictures such as Figure 1.1(l) from which it is not absolutely clear how the pattern is repeated.

1.2 The parts of a tiling

In a tiling, the tile boundaries are just as important as the tiles themselves — indeed, from the point of view of drawing them, they are all-important, as they are what we actually draw! In fact, we can characterize a tiling by its tile boundaries, that is, by its *net*.

> *Definition 1.2 Net of a tiling*
>
> The **net** of a tiling is the set of all points in the plane that are boundary points of tiles.

Exercise 1.4

Why must any point in the net of a tiling be a boundary point of *at least two* tiles?

The result of Exercise 1.4 shows that an alternative way of defining the net of a tiling would be to say that it is the set of all points in the plane that belong to *more than one* tile.

You may wonder whether the net of a tiling consists entirely of points belonging to *exactly two* tiles. The answer is *no*, for the following reason. Consider walking round the boundary of some (shaded) tile in a tiling. There are two possible cases that might occur, illustrated in Figure 1.5 (where the tile we walk round is shaded in each case). Figure 1.5(a) illustrates the case where there is only one tile surrounding the shaded tile, while Figure 1.5(b) illustrates the case where there are more than one. In the first case, the tile surrounding the shaded tile cannot be a topological disc, as the shaded tile would have to constitute a hole in the surrounding tile; thus this is not an example of a tiling as we have defined it. In the second case, there have to be points on the boundary of the shaded tile where one surrounding tile takes over from another; and these points must belong to at least three tiles. (In the figure, two of these points belongs to three tiles, one to four and the other to five.)

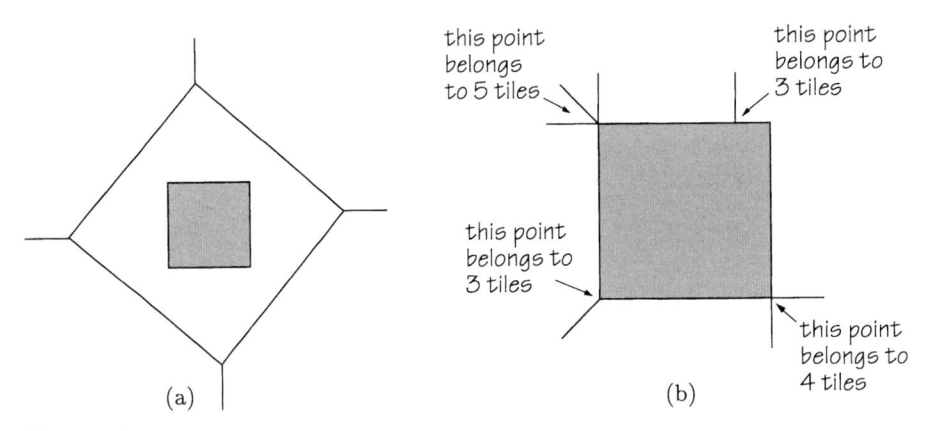

Figure 1.5

The points on the boundaries of more than two tiles are clearly the meeting points of more than two boundary *lines* of the tiling. Indeed, the net of a tiling can be regarded as a set of points in the plane (the points on the boundaries of three or more tiles), joined by a set of (straight or curved) lines, which are the boundaries between exactly two tiles. In Figure 1.6, we depict a portion of a tiling and we emphasize those points that are on three or more tiles by drawing them as large dots; this makes it clear that the net of the tiling consists of lines joining these dots.

These two types of boundary point are of fundamental importance and require formal definition.

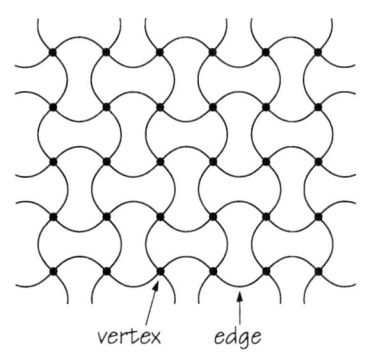

Figure 1.6

Definition 1.3 Vertices and edges of a tiling

The **vertices** of a tiling (or of the net of a tiling) are the points on the boundary of more than two tiles. The **edges** are the lines (curved or straight) which join the vertices and consist of points on the boundary of exactly two tiles.

If you have studied graph theory, you may be struck by the analogy with the concepts of vertices and edges of a *graph*. Indeed, the net of a tiling is an *infinite graph* in the language of graph theory.

Thus the vertices and edges of a tiling, as well as the tiles themselves, are important component parts. In fact, these three components of a tiling are so important that it is worth defining a collective name for them.

> **Definition 1.4 Parts of a tiling**
>
> The tiles, edges and vertices of a tiling, taken together, are called the **parts** of a tiling.

Often we need to refer to particular parts of a tiling (and to the tiling as a whole, for that matter). In this course, we shall use a script upper-case letter (often $\mathcal{T}$) to refer to a tiling, and ordinary upper-case letters to refer to its parts. Usually, the tiles to which we wish to refer are labelled T_1, T_2, etc., the edges E_1, E_2, etc., and the vertices V_1, V_2, etc. For example, in Figure 1.7 (in which not all the parts are labelled), the edge E_3 joins the vertices V_1 and V_2, and marks the boundary between tiles T_1 and T_4.

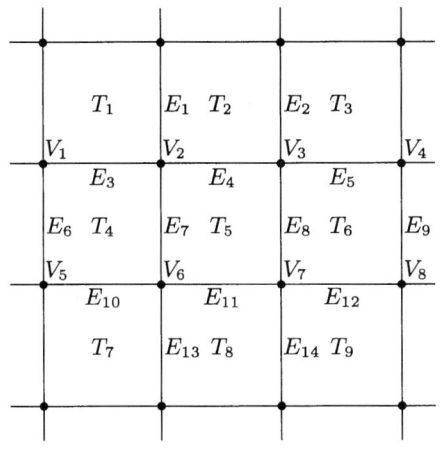

Figure 1.7

Exercise 1.5

Referring to Figure 1.7, name:

(a) the vertices and edges on the boundary of T_5;

(b) the tiles of which E_8 lies on the boundary;

(c) the tiles of which V_3 lies on the boundary.

1.3 Adjacency and incidence

We need a concise way of talking about the relations between parts of a tiling. We now introduce two such relations: *adjacency* and *incidence*.

Adjacency is a relation that may exist between two parts of the same type (two tiles, two edges or two vertices). As you might expect, adjacency means 'being next to'. Thus we shall take two tiles to be adjacent if they are separated by an edge; for example, in Figure 1.7, tiles T_4 and T_5 are adjacent. However, tiles T_2 and T_6 are *not* adjacent as far as this course is concerned, despite the fact that they share a common *vertex*.

In some contexts, such as image processing, tiles sharing a common vertex are regarded as adjacent, but as far as we are concerned the shared boundary *must be an edge*.

What about the *vertices* of a tiling? The most obvious definition of 'being next to' in this case is that two vertices are next to each other if there is an edge joining one to the other, and this is the definition of vertex-adjacency that we shall use. For example, in Figure 1.7 the vertex V_2 is adjacent to V_3 and V_6 (among others), but is not adjacent to V_4 or V_7.

As far as the *edges* of a tiling are concerned, 'being next to' can be interpreted in two ways. As we trace round a particular tile, the edges occur in a definite order, so we can regard two edges as adjacent if one occurs immediately after the other on going round some tile (in either direction). Thus, for example, as we go clockwise round T_5 in Figure 1.7, starting at edge E_4, the edges occur in the order E_4, E_8, E_{11}, E_7, so that E_4 is adjacent to E_7 and E_8 but not to E_{11}. Similarly, if we look at the way the edges radiate from a vertex, they occur in a definite order, so we can regard two edges as adjacent if one occurs immediately after another on going round some vertex. Going clockwise round V_2, for example, starting at edge E_1, we have E_1, E_4, E_7, E_3, so that E_4 is adjacent to E_1 and E_7 but not to E_3. Which definition shall we take? In fact, it does not matter — two edges that are next to each other round a tile are also next to each other round some vertex, and vice versa. Another way of putting it is to say that two edges are adjacent if they *both* bound a common tile *and* have a common vertex.

Going round T_2, the other tile bounded by E_4, shows that E_4 is also adjacent to E_1 and E_2.

Going round V_3, the other vertex of E_4, shows that E_4 is also adjacent to E_2 and E_8.

Definition 1.5 Adjacency

Two distinct tiles are **adjacent** if they share a common edge. Two distinct vertices are **adjacent** if they are joined by an edge. Two distinct edges are **adjacent** if they share a common vertex and bound a common tile.

If you have studied graph theory, you should note that the definition of adjacency between edges here is different from the graph-theoretic definition, in which two edges which *share a vertex* are regarded as adjacent. Thus, in Figure 1.7, E_4 is adjacent to E_3 by the graph-theoretic definition but *not* by the M336 definition.

Exercise 1.6

Consider again the tiling depicted in Figure 1.7, and answer the following questions:

(a) Which tiles are adjacent to T_5?

(b) Are tiles T_6 and T_8 adjacent? If so, why? If not, why not?

(c) Are edges E_7 and E_{13} adjacent? If so, why? If not, why not?

Exercise 1.7

In each of the tilings in Figure 1.8, how many edges are adjacent to any particular edge of the tiling?

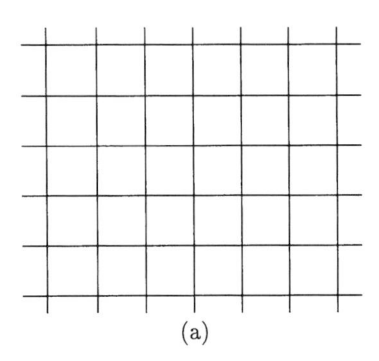

Figure 1.8

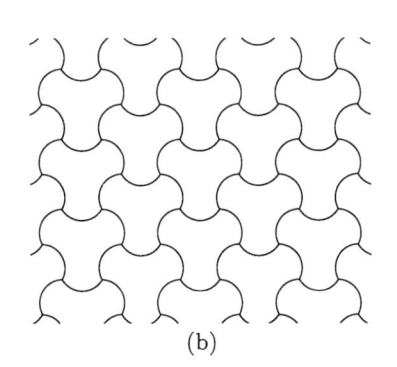

The *incidence* relation is somewhat similar to the adjacency relation except that, instead of relating parts of the *same* type, it relates parts of *different* types. Thus there are three possible types of incidence: between tiles and edges, between tiles and vertices, and between edges and vertices.

'Incident' in this context means 'touching' or 'being involved with'. Thus the edges and vertices that are incident with a particular tile T in a tiling are just those that lie on the boundary of T. Similarly, any edge E is bounded by the two vertices that lie at each end of it; and these are the vertices that are incident with E.

> **Definition 1.6 Incidence**
>
> The edges and vertices on the boundary of a tile T in a tiling are **incident** with T. Similarly, T is **incident** with the edges and vertices on its boundary. Also, if an edge E joins vertices V_1 and V_2, then E is **incident** with V_1 and V_2, and these vertices are **incident** with E.

Remember: two parts of the *same* type may or may not be *adjacent*, whereas two parts of *different* types may or may not be *incident*.

These definitions may seem a bit of a mouthful, but with a little practice you should find them quite natural.

Exercise 1.8

In Figure 1.7, which vertices and which tiles are incident with E_4?

Another basic concept is needed before we leave this section. In any tiling, it is useful to know *how many tiles are adjacent to a given tile*, and also *how many vertices are adjacent to a given vertex*. These numbers are called the *degrees* of the corresponding parts.

> **Definition 1.7 Degree of a tile or vertex**
>
> The **degree** of a tile (or vertex) in a tiling is the number of other tiles (or vertices) to which it is adjacent.

As each edge is always adjacent to *exactly four* other edges, there is no point in defining the 'degree' of an *edge* of a tiling.

For example, in the tiling depicted in Figure 1.7, the tiles and the vertices all have degree 4. However, it is perfectly possible for the degrees of the tiles and/or of the vertices to vary in any particular tiling.

Exercise 1.9

Consider the tiling in Figure 1.9. What are the degrees of the vertices and of the tiles?

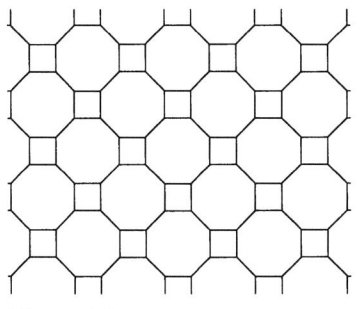

Figure 1.9

The definitions in this section apply to every tiling which you will meet in this course, but there is very little that can be proved mathematically at such a level of generality. We cannot hope, for example, to classify in any meaningful way all the tilings which obey the very general definition on page 10 (Definition 1.1). To find results that are mathematically interesting, we must set our sights a little lower and define certain special types of tiling. The simplest way to do this is to impose restrictions on the shapes of the individual tiles; this is done in the next section.

2 TILINGS USING POLYGONS

Before we discuss tilings using polygons, it is worth giving a formal definition of the familar concept of a polygon.

> *Definition 2.1 Polygon*
>
> A **polygon** is a topological disc bounded by a finite set of straight-line segments, its **sides**. Each point where two sides meet is a **corner** of the polygon.

2.1 Polygonal tilings

A polygonal tiling is very easy to define.

> *Definition 2.2 Polygonal tiling*
>
> A **polygonal tiling** is one in which each tile is a polygon.

Exercise 2.1

Which of the tilings depicted in Figure 1.1 are polygonal?

Remember, from Exercise 1.3, that only (a)–(g) and (l) in Figure 1.1 depict tilings.

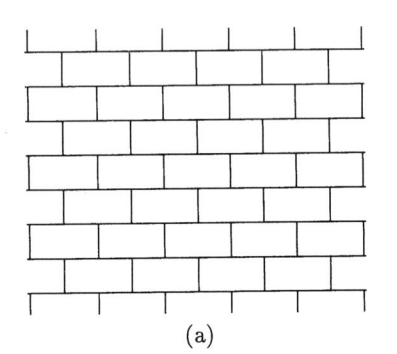

A familiar polygonal tiling is the *brick-wall tiling*, shown in Figure 2.1(a). Notice that although the face of each brick has four sides, each brick is adjacent to six other bricks. Thus, considered as a tiling, each tile has six edges, not four. Each horizontal *side* of the face of each brick constitutes *two edges* of the tiling, as an edge joins just two vertices. Furthermore, although the face of each brick has *four corners*, when considered as a tiling each tile has *six vertices*. Figure 2.1(b) emphasizes these points by showing the vertices as large dots.

Builders call this particular way of laying bricks *stretcher-bond*.

Figure 2.1

On the other hand, Figure 2.2(a) shows another polygonal tiling, in which each *edge* of a tile constitutes *two sides* of the corresponding (non-convex) polygon. Moreover, each tile, when considered as a polygon, has *eight corners*, only *four* of which are *vertices* of the tiling. Figure 2.2(b) emphasizes these points by showing the vertices as large dots.

A polygon is *convex* if each of its internal angles is less than π (that is, 180°), or equivalently, if for every pair of distinct points in the polygon, the straight-line segment joining them lies wholly in the polygon. Each polygon in Figure 2.2 has two of its internal angles greater than π and hence is *non-convex*.

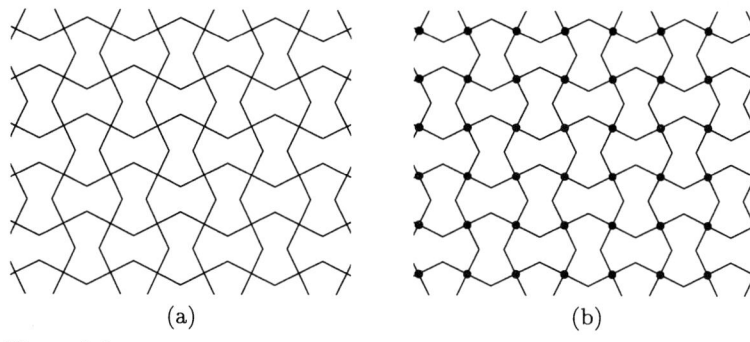

Figure 2.2

Warning The examples in Figures 2.1 and 2.2 show that the concepts of *edges* and *vertices* of a *tiling* are different from the concepts of *sides* and *corners* of a *polygon*, and that it is important to distinguish these carefully when we are talking about polygonal tilings.

Exercise 2.2

Draw a portion of a polygonal tiling in which there are some sides of polygons that constitute more than one edge of the tiling *as well as* some edges of the tiling that constitute more than one side of some of the polygons.

Polygonal tilings are almost as hard to classify as tilings in general, if we allow edges to consist of any number of sides and vice versa. After all, given *any* tiling, we can convert it into a polygonal tiling by taking each curved edge and replacing the curve with an approximation consisting of a number of short, straight-line segments, as Figure 2.3 illustrates. We would like to exclude this kind of case, and we do so by the following definition.

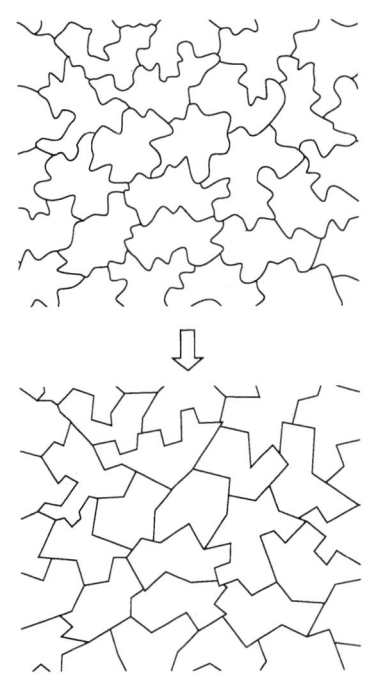

Figure 2.3 Approximating tiling (f) from Figure 1.1 by a polygonal tiling.

Definition 2.3 Edge-to-edge tiling

A polygonal tiling $\mathcal{T}$ is an **edge-to-edge tiling** if each side of each polygon corresponds to exactly one edge of $\mathcal{T}$, and vice versa, so that the edges and the sides of $\mathcal{T}$ are identical.

A consequence of this definition is that the internal angle at a vertex of a tile in an edge-to-edge tiling cannot be exactly π, though it may be less than or greater than π.

We can immediately deduce from Figures 2.1 and 2.2 that neither of these polygonal tilings is an edge-to-edge tiling, since in Figure 2.1 each horizontal side of each polygon corresponds to *two* edges of each tile and since in Figure 2.2 each edge of each tile corresponds to *two* sides of each polygon. Similarly, in Figure 2.4, we can deduce that tiling (a) is edge-to-edge but that tiling (b) is not. In (a), there is an obvious one-to-one correspondence between sides and edges. In (b), each edge of the tiling separating a star shape from a small hexagon corresponds to *two* sides of each of these shapes, while each side of each large hexagon corresponds to *portions* of *two* edges of the tiling.

Notice, in Figure 2.1, that the internal angle is exactly π at the vertices at the centres of the horizontal sides of the polygons. Notice how all such vertices must divide a side of a polygon into two or more edges of a tiling, and hence must disqualify such tilings from being edge-to-edge tilings.

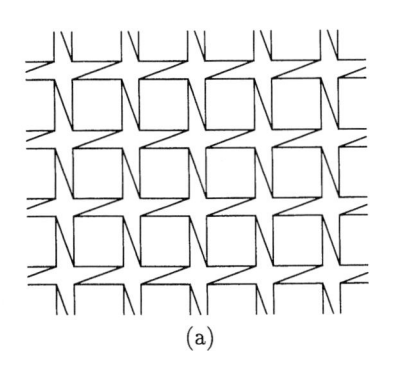

Figure 2.4

Exercise 2.3

Which of the tilings in Figure 2.5 are edge-to-edge?

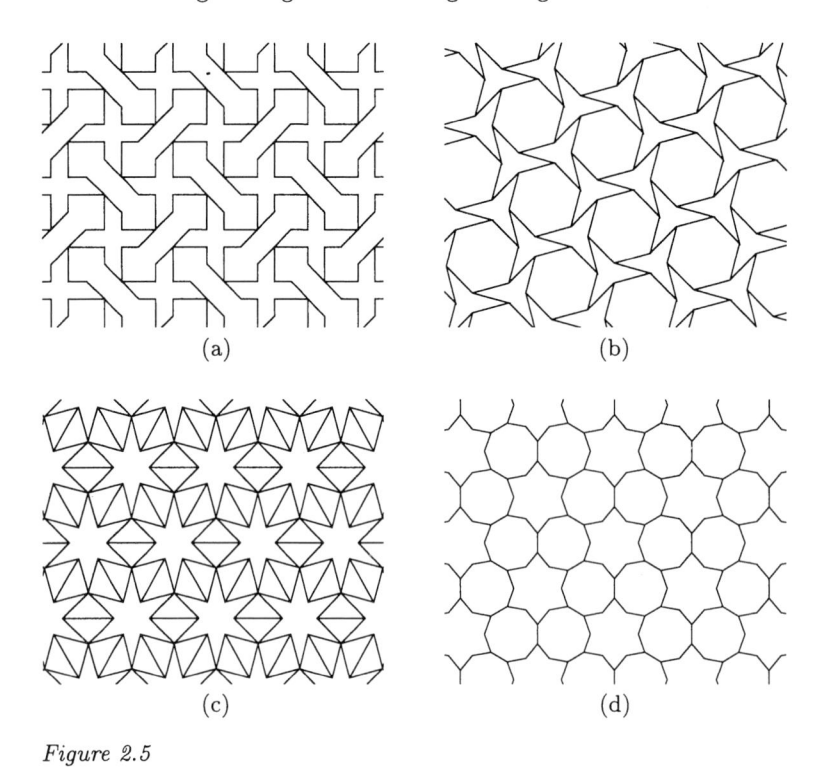

Figure 2.5

The above examples have used polygons with a wide variety of shapes, so it should not surprise you to learn that even edge-to-edge tilings are so varied as to defy classification. To obtain some sort of classification, we need to restrict our range of tilings even further. One way to do this is to insist that each tile is a **regular polygon** — that is, a polygon all of whose sides and angles are equal.

If a tiling is required both to be edge-to-edge and to consist of regular polygons (not necessarily all the same), this reduces the field quite considerably, though the number of possibilities is still surprisingly large. We deal with tilings of this kind next.

2.2 The Archimedean tilings

There are three particularly familiar edge-to-edge tilings using regular polygons: those using only equilateral triangles, those using only squares and those using only regular hexagons. These three tilings are the **regular tilings**, and are so important that we give them names. We shall call them $\mathcal{R}_3$, $\mathcal{R}_4$ and $\mathcal{R}_6$, respectively.

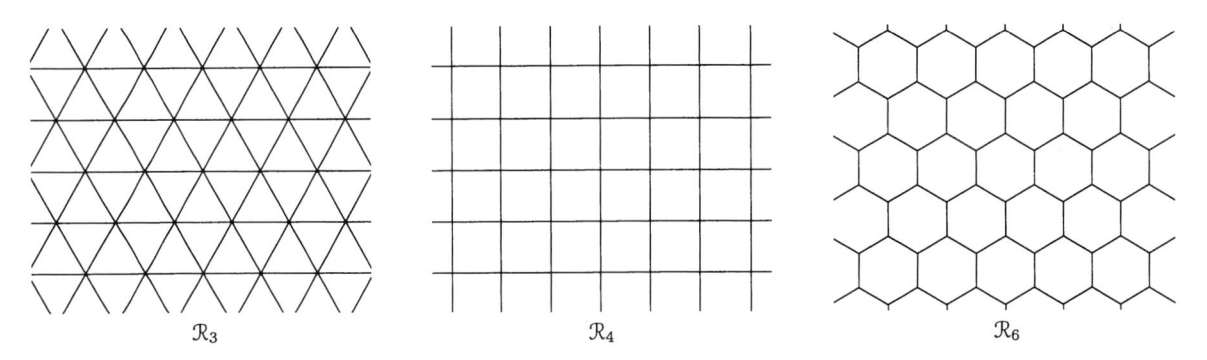

Figure 2.6   The regular tilings $\mathcal{R}_3$, $\mathcal{R}_4$ and $\mathcal{R}_6$.

It may seem at first sight that these are the *only* edge-to-edge tilings by regular polygons; indeed, if we *also* insist that *all the tiles in a tiling be congruent*, then this is true. This result is important enough to be given a name.

> *Theorem 2.1 Regular tiling theorem*
>
> The regular tilings $\mathcal{R}_3$, $\mathcal{R}_4$ and $\mathcal{R}_6$ are the only edge-to-edge tilings by congruent regular polygons.

If you are not familiar with the geometric concept of *congruence*, see Definition 4.1 on page 32. In the context of regular polygons, two such polygons are congruent if they have the same number of sides and those of one are the same length as those of the other.

Proof

Let $\mathcal{T}$ be an edge-to-edge tiling by congruent regular polygons. The angles at each vertex must be equal, and their sum must be 2π. But the only regular polygons having internal angles that divide 2π are the triangle (internal angle $\pi/3$, or 60°), the square (internal angle $\pi/2$, or 90°), and the hexagon (internal angle $2\pi/3$, or 120°). The internal angle of a regular pentagon is $3\pi/5$, which does not divide 2π; and for anything with more sides than a hexagon, the internal angle is greater than $2\pi/3$ but less than π, and so cannot divide 2π. Thus the tilings $\mathcal{R}_3$, $\mathcal{R}_4$ and $\mathcal{R}_6$ are the only edge-to-edge tilings by congruent regular polygons. ∎

We say that a number x *divides* a number y if $z = y/x$ is an integer.

If we remove the condition that all the polygons be congruent, but still insist that they all be regular polygons, fitted edge-to-edge, then many more possibilities arise.

At this stage, you should open your Geometry Envelope. You will find several cards, printed on both sides, and also several transparent overlays to go with certain cards. For the moment, please ignore the overlays, and all the cards except Tiling Cards 1 and 2.

Look now at the eleven pictures on Tiling Card 1 (five on Side 1 and six on Side 2). The first three pictures on Tiling Card 1 are of the regular tilings depicted in Figure 2.6, and the other eight are edge-to-edge tilings by a mixture of regular polygons. These eight are called the **semi-regular tilings**. The semi-regular tilings, together with the regular tilings, $\mathcal{R}_3$, $\mathcal{R}_4$ and $\mathcal{R}_6$, are collectively called the **Archimedean tilings**.

This set of eleven tilings does not by any means exhaust the possible edge-to-edge tilings by regular polygons. For example, six more such tilings are depicted on Side 1 of Tiling Card 2.

Only four of the cards in the *Geometry Envelope*, namely Tiling Cards 1, 2 and 3 (and some of the corresponding overlays) and the Isometry Toolkit card, are relevant to this unit; the others will be used in conjunction with later Geometry units. The significance of the bracketed sets of figures under some of the tilings will be explained later in this section.

Archimedes (287(?)–212 BC) investigated those polygonal solids whose faces are regular polygons. He probably did not investigate polygonal tilings as such; nevertheless, the name 'Archimedean tilings' is in common usage.

Exercise 2.4

Using the fact that a regular hexagon can be divided into six equilateral triangles, show that there are infinitely many different edge-to-edge tilings by regular polygons.

It is probably not clear to you why the tilings of Tiling Card 1 are called Archimedean while those of Tiling Card 2, Side 1 are not. The reason is that there is in fact a mathematically significant difference between them. It is that, in the Archimedean tilings, every vertex is 'the same' as every other, in the sense of having the same arrangement of polygons (tiles) incident with it. This arrangement is reflected in the bracketed symbol which appears under each tiling on Tiling Card 1. It is called the *vertex type* for the tile.

For example, the fourth tiling on Side 1 of Tiling Card 1 has the vertex type $(3, 3, 3, 3, 6)$, because the tiles incident with any vertex are four triangles and a hexagon (four tiles of degree 3 and one of degree 6). The second tiling on Side 2 of Tiling Card 1 has the vertex type $(3, 4, 6, 4)$, because the tiles incident with any vertex of that tiling are of degrees $3, 4, 6, 4$ *in that order* (provided we start at the tile of degree 3).

Note that the fifth tiling on Side 2 of Tiling Card 1, with the vertex type $(4, 6, 12)$, has some vertices at which the tiles of degrees $4, 6, 12$ occur in *clockwise* order and some at which they occur in *anticlockwise* order. It is a moot point whether we regard these as the same or as different vertex types. If they are to be regarded as different, then this tiling cannot be regarded as an Archimedean tiling, as we have two different vertex types. Most mathematicians (including the M336 course team) regard these as the same, and so allow this tiling as an Archimedean tiling. (Notice that, for the other ten Archimedean tilings, the order of the degrees of the tiles surrounding a vertex is the same — subject to appropriate starting tile — irrespective of whether you move clockwise or anticlockwise around the vertex.)

If we were to start at a different tile, we would get a different symbol for the vertex type. Thus, $(4, 6, 4, 3)$, $(6, 4, 3, 4)$ and $(4, 3, 4, 6)$ are all just as valid as symbols for the vertex type of the second tiling on Side 2 of Tiling Card 1 as is $(3, 4, 6, 4)$. All four symbols have exactly the same meaning in this context. In fact, all eight semi-regular tilings have several equivalent symbols for their vertex type, depending on the starting tile used. The symbols shown on Tiling Card 1, however, are the standard ones and we shall use these in this course.

Exercise 2.5

Each of the tilings on Side 1 of Tiling Card 2 has vertices of two different types. For each of the first two tilings (i.e. the two in the top row), list both the vertex types that are present.

We now formalize the concepts of *vertex type*, *Archimedean tiling* and *regularity* of a tiling.

Definition 2.4 Vertex type

Let V be any vertex in a tiling $\mathcal{T}$. The **vertex type** of V is given by listing the degrees of the tiles incident with V, starting with any one of them and proceeding in either clockwise or anticlockwise order. This list is conventionally placed in parentheses.

Definition 2.5 Equality of vertex types

Two vertex types $(a_1, a_2, \ldots, a_n)$ and $(b_1, b_2, \ldots, b_n)$ are the **same** if, for some i $(0 \leq i \leq n-1)$, either $a_j = b_{i+j}$ $(j = 1, \ldots, n)$ or $a_j = b_{i-j}$ $(j = 1, \ldots, n)$, where we interpret the subscripts modulo n.

Definition 2.6 Vertex-uniform tiling

A tiling is **vertex-uniform** if all its vertex types are the same.

Definition 2.7 Archimedean, regular and semi-regular tilings

A tiling is **Archimedean** if it is edge-to-edge and vertex-uniform, and all the tiles are regular polygons. If, in addition, all the tiles are congruent, then the tiling is **regular**; otherwise it is **semi-regular**.

Thus, the eight semi-regular tilings are vertex-uniform but have a mixture of tile degrees, while the three regular tilings are vertex-uniform and consist of tiles all of the same degree.

Exercise 2.6

What are the vertex types of the three regular tilings?

The result that the eleven tilings on Tiling Card 1 are indeed the only Archimedean tilings is an important one; it is stated below in the form of a theorem.

> **Theorem 2.2 Archimedean tiling theorem**
>
> The only vertex types which give rise to Archimedean tilings are the following:
>
> $(3,3,3,3,3,3); (4,4,4,4); (6,6,6); (3,3,3,3,6); (3,3,3,4,4);$
>
> $(3,3,4,3,4); (3,4,6,4); (3,6,3,6); (3,12,12); (4,6,12); (4,8,8).$

Sketch of proof

The proof is not conceptually difficult, but it is tedious. We consider the angles which can be internal angles of regular polygons, and construct a list of possible ways in which these angles can be fitted together to make an angle of 2π. This gives all possible vertex types involving only regular polygons. There are in fact 21 of these vertex types. There is no need to remember them, but as a matter of interest they are the above eleven vertex types, together with the following ten:

$(3,3,4,12); (3,4,3,12); (3,3,6,6); (3,4,4,6); (3,7,42);$

$(3,8,24); (3,9,18); (3,10,15); (4,5,20); (5,5,10).$

This sketch proof is optional.

The next step is to try to construct a tiling based on each of these types. Eleven of the types give a pattern which is vertex-uniform, and these are the eleven Archimedean tilings. The other ten simply break down. For example, consider the vertex type $(3,7,42)$. Let E_1, E_2, E_3 be the edges of some triangular tile T in such a tiling. If E_1 is the boundary between T and a 42-gon, then E_2 and E_3 must be the boundary edges between T and two heptagons (i.e. 7-gons) (Figure 2.7). Now look at the vertex of T incident with E_2 and E_3. It has two heptagons and a triangle incident with it, so it cannot have vertex type $(3,7,42)$. (Indeed, the angle left over is too small to accommodate any regular polygon whatever!) ∎

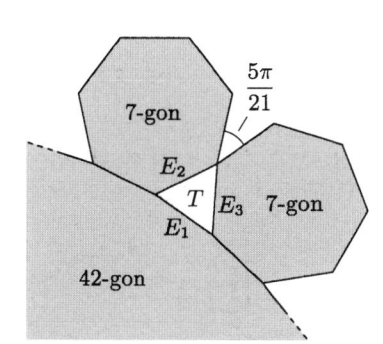

■ *Figure 2.7*

Exercise 2.7

Three of the Archimedean tilings can be drawn by drawing sets of parallel infinite straight lines. Which are they?

Before we leave the topic of the Archimedean tilings, we should mention one rather unexpected feature of the first of the semi-regular tilings, $(3,3,3,3,6)$. There are two pictures of this tiling in Figure 2.8. These pictures are mirror images of each other; and it is impossible to superimpose one on the other without turning one of them over. This is in fact the only Archimedean tiling for which it is true that the mirror image cannot be superimposed on the original tiling. You can verify this by using the two transparent overlays labelled Overlay 1 that go with Tiling Card 1: for each tiling except $(3,3,3,3,6)$, the overlay picture will fit exactly over the card picture whichever side of the overlay is uppermost; but in the case of $(3,3,3,3,6)$, if the overlay is turned over, to produce the mirror image, then the overlay picture will *not* fit over the card picture.

The full titles of these two overlays are Tiling Card 1, Side 1, Overlay 1 and Tiling Card 1, Side 2, Overlay 1.

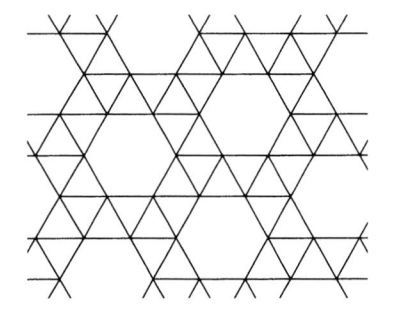

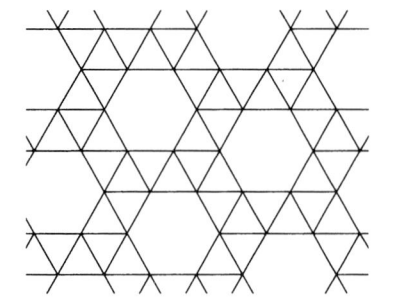

Figure 2.8

It would be possible to regard the two pictures in Figure 2.8 as representing two different tilings. But, just as mathematicians regard the vertex types $(4, 6, 12)$ and $(4, 12, 6)$ as the same because one is (in a sense) the mirror image of the other, so it is normal to regard these two versions of the tiling $(3, 3, 3, 3, 6)$ as the same.

The idea of superimposing mirror images leads on to the next section. In Section 3, we revise and extend the techniques for recognizing and manipulating operations such as reflections, rotations and more general transformations of the plane.

You may have met these techniques in your previous studies.

3 AFFINE TRANSFORMATIONS AND ISOMETRIES

Much of the fascination of tilings lies in their intricate *symmetries*. That is to say, there are many ways in which one can pick up the whole pattern and shift or rotate it or turn it over without disturbing its appearance. In other words, if you have two copies of the pattern, one printed on transparent material, then there are many different ways in which you can shift, rotate or turn over the transparent copy and then place it on top of the other copy, so that the two match each other exactly.

What if you pick up a copy of the pattern and put it down in such a way that it does *not* fit exactly over the original? The original and displaced copies of the pattern are still related by a transformation of the plane — one in which the distance between any pair of points remains constant. Any such function from the plane to itself is called an *isometry*. In fact, the concept of an isometry does not in itself involve any notion of a pattern imposed on the plane. However, those particular isometries that map a pattern drawn in the plane exactly onto itself are called *symmetries* of that pattern.

More generally, we can start with a given tiling and map it to a possibly different-looking tiling by changing the scales in the x- and y-directions, possibly skewing the axes as well. As you will see in Section 4, we often obtain tilings with new properties in this way, so such operations are of interest to us. Such maps are called *affine transformations*.

In Subsection 3.1 we remind you how to set up a coordinate system for the plane, and how to characterize a *linear* transformation by means of a *matrix* once such a system has been set up. In Subsection 3.2 we discuss the general form of an *affine* transformation of the plane, and in Subsections 3.3 and 3.4 we take a close look at the subclass of affine transformations that constitute the *isometries*.

If you have recently studied a second-level pure or applied mathematics course, you may wish to skip Subsection 3.1.

3.1 Coordinate systems and linear transformations

You may remember from your previous studies that, if we wish to work algebraically with transformations of the plane (or three-dimensional space), we require a way of modelling the plane (or three-dimensional space) in algebraic terms. To do this for the plane, we make the following constructions:

(a) We choose a point O in the plane to be the *origin*.

(b) We draw a line through O and call it the *x-axis*, then we draw a second line through O at right angles to the first and call it the *y-axis*.

(c) We choose a unit length and use it to measure distances between points.

Every point P in the plane now has an *x-coordinate* and a *y-coordinate*, which can be found by dropping perpendiculars to the x-axis and y-axis respectively. (The origin has x-coordinate 0 and y-coordinate 0.) If the x-coordinate is p and the y-coordinate is q, then the ordered pair (p, q) is the two-dimensional *vector* representing the point P in this coordinate system. It is conveniently denoted by a single bold lower-case letter, such as **p**.

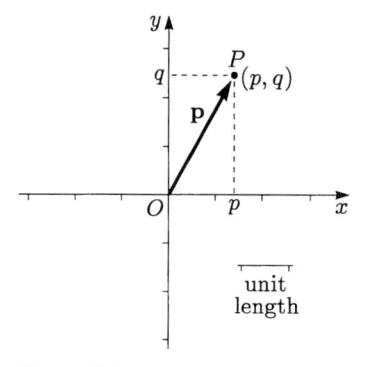

Figure 3.1

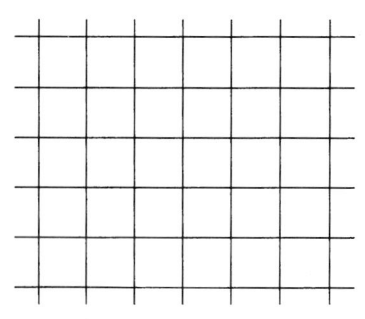

An important point to note is that each of the constructions (a), (b) and (c) above introduces an element of choice, and each choice has an effect on the coordinates that are actually obtained for a point P. In some respects this is a pity, since it means that in order to specify a transformation (or even a single point) in algebraic terms, we need to specify which choices we have made under (a), (b) and (c) above. In other respects it is an advantage, as there may well be a particularly convenient choice that can be made. For example, if we were discussing the tiling in Figure 3.2, it would make sense to choose the origin to be at one of the vertices, the axes to be parallel with the tile edges, and the unit length to be the length of any one of the edges.

Figure 3.2

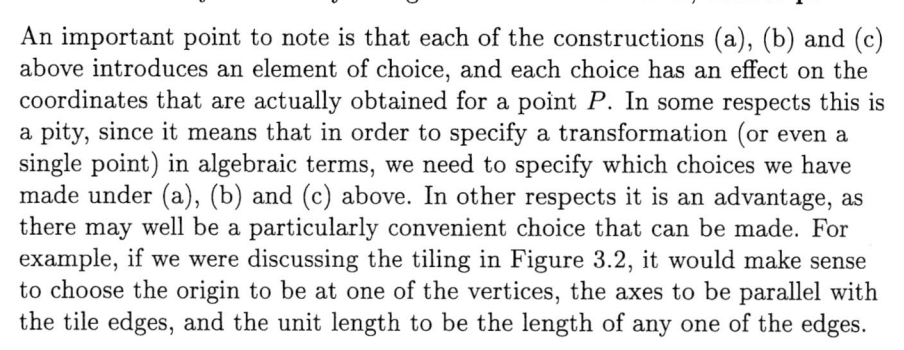

Once a particular coordinate system has been chosen, we have the important concept of a **linear transformation** on $\mathbb{R}^2$. This is a mapping λ from $\mathbb{R}^2$ to $\mathbb{R}^2$ with the property that, once we know the images under λ of the *standard basis vectors* $(1,0)$ and $(0,1)$ of $\mathbb{R}^2$, i.e. once we know $\lambda(1,0)$ and $\lambda(0,1)$, we can calculate the image $\lambda(x,y)$ of any vector (x,y) in the plane, as follows:

$$\lambda(x,y) = \lambda(x(1,0) + y(0,1)) = x\lambda(1,0) + y\lambda(0,1).$$

That is to say, if $\lambda(1,0)$ is the vector (a,b) and $\lambda(0,1)$ is the vector (c,d), then

$$\lambda(x,y) = (ax + cy, bx + dy). \tag{3.1}$$

Writing (x,y) and $(ax+cy, bx+dy)$ as columns, we can express the relationship between them by means of matrix multiplication, using the 2×2 matrix

$$\mathbf{A} = \begin{bmatrix} a & c \\ b & d \end{bmatrix}.$$

We have

$$\begin{bmatrix} ax + cy \\ bx + dy \end{bmatrix} = \begin{bmatrix} a & c \\ b & d \end{bmatrix} \begin{bmatrix} x \\ y \end{bmatrix}.$$

This matrix **A** is known as the **matrix of the transformation** and the transformation is often denoted by $\lambda[\mathbf{A}]$. Such a transformation is *invertible* if it has an inverse (that is, if every point in the plane is the image of exactly one point), and this is true if and only if **A** has non-zero determinant. The inverse transformation $(\lambda[\mathbf{A}])^{-1}$ has transformation matrix $\mathbf{A}^{-1}$, i.e. $(\lambda[\mathbf{A}])^{-1} = \lambda[\mathbf{A}^{-1}]$.

The symbol $\mathbb{R}$ is widely used to denote the real number system. Since the above model represents points as pairs of numbers in $\mathbb{R}$, we use the notation $\mathbb{R}^2$ to denote the plane. Similarly, three-dimensional space is denoted by $\mathbb{R}^3$.

Do not confuse, for example, the notation $\lambda(1,0)$, which represents the image of $(1,0)$ under λ, with the notation $x(1,0)$, which represents the multiplication of the vector $(1,0)$ by the scalar x.

When you are making notes, answering assignment questions, etc., you can just write R for $\mathbb{R}$, unless you are using R for something else. You are also advised to write the matrix **A** as $\underline{A}$ and the vector **p** as $\underline{p}$, and generally to denote any bold-face character in a mathematical expression by underlining it.

We can if we wish use invertible linear transformations to produce new tilings from old, but they have a disadvantage. A linear transformation must map the origin to the origin, as may be deduced from Equation 3.1 with $(x, y) = (0, 0)$. Now, for any transformation other than the identity transformation, there will be plenty of points that are *not* mapped to themselves. So, if such a point is chosen as the origin, the transformation is not linear. Thus the choice of origin for the coordinate system determines whether or not a particular transformation is linear.

The more general concept of an *affine transformation* does not suffer from this drawback; as we shall see, a change of coordinate system does not affect the question of whether or not a particular transformation is affine. We shall now define this concept.

> You should have met affine transformations before, in your previous studies.

3.2 Affine transformations

Definition 3.1 Affine transformation

An **affine transformation** of $\mathbb{R}^2$ (or of $\mathbb{R}^n$) is a transformation of the form

$$\mathbf{x} \mapsto \mathbf{A}\mathbf{x} + \mathbf{p},$$

where $\mathbf{A}$ is the matrix of an invertible linear transformation and $\mathbf{p}$ is some constant vector.

In coordinate terms (restricting our attention to the plane), if

$$\mathbf{A} = \begin{bmatrix} a & c \\ b & d \end{bmatrix}, \quad \mathbf{x} = \begin{bmatrix} x \\ y \end{bmatrix}, \quad \mathbf{p} = \begin{bmatrix} p \\ q \end{bmatrix},$$

then the transformation maps the point (x, y) to the point

$$(ax + cy + p, bx + dy + q).$$

> The column and row forms for writing out vector coordinates will be used interchangeably in this course.

We can think of such a transformation as being accomplished in two stages: first, map $\mathbf{x}$ to $\mathbf{A}\mathbf{x}$ by a linear transformation, then map $\mathbf{A}\mathbf{x}$ to $\mathbf{A}\mathbf{x} + \mathbf{p}$ by a *translation* — a transformation which simply shifts the plane bodily without altering its orientation. Writing $f : \mathbb{R}^2 \to \mathbb{R}^2$ for the affine transformation as a whole, we have

$$f = t[\mathbf{p}] \circ \lambda[\mathbf{A}];$$

that is, f is the composite of the two functions $\lambda[\mathbf{A}]$ (the **linear part** of f) and $t[\mathbf{p}]$ (the **translation part**), where $\lambda[\mathbf{A}]$ is the linear transformation with matrix $\mathbf{A}$ and $t[\mathbf{p}]$ is translation by the vector $\mathbf{p}$.

> Recall that we always compose functions from right to left.
>
> Notice that $(t[\mathbf{p}])^{-1} = t[-\mathbf{p}]$.

Exercise 3.1

Let $f : \mathbf{x} \mapsto \mathbf{A}\mathbf{x} + \mathbf{p}$ and $g : \mathbf{x} \mapsto \mathbf{B}\mathbf{x} + \mathbf{q}$ be two affine transformations. Show that the composite transformation $g \circ f$ is also affine. Write the linear and translation parts of the composite transformation in the compact notation (i.e. using t and λ) developed above.

It is easy to compose affine transformations using the compact notation. This can be done using three simple rules.

Rules for composing translations and linear transformations

Let $\mathbf{p}, \mathbf{q}$ be any two vectors and let $\mathbf{A}, \mathbf{B}$ be any two invertible matrices.

Rule 1: $t[\mathbf{p}] \circ t[\mathbf{q}] = t[\mathbf{p} + \mathbf{q}]$.

Rule 2: $\lambda[\mathbf{A}] \circ \lambda[\mathbf{B}] = \lambda[\mathbf{AB}]$.

Rule 3: $\lambda[\mathbf{A}] \circ t[\mathbf{p}] = t[\mathbf{Ap}] \circ \lambda[\mathbf{A}]$.

Rule 1 simply says that translation by one vector, then by another, is the same as translation by the sum of the two vectors.

Rule 2 is the familiar property of matrix multiplication: $(\mathbf{AB})\mathbf{x} = \mathbf{A}(\mathbf{Bx})$.

Rule 3 is a little less obvious, but is easy to check. For any vector $\mathbf{x}$, the result of applying $\lambda[\mathbf{A}] \circ t[\mathbf{p}]$ to $\mathbf{x}$ is as follows: applying $t[\mathbf{p}]$ to $\mathbf{x}$ results in $\mathbf{x} + \mathbf{p}$, then applying $\lambda[\mathbf{A}]$ to $\mathbf{x} + \mathbf{p}$ gives $\mathbf{A}(\mathbf{x} + \mathbf{p}) = \mathbf{Ax} + \mathbf{Ap}$; and this is the same result as first applying $\lambda[\mathbf{A}]$ to $\mathbf{x}$, to give $\mathbf{Ax}$, and then applying $t[\mathbf{Ap}]$ to $\mathbf{Ax}$. In other words, each side simply describes the same affine transformation $f : \mathbf{x} \mapsto \mathbf{Ax} + \mathbf{Ap}$, and thus the rule holds.

Using these three rules, we can derive two further rules for composing and finding inverses of affine transformations given in terms of linear and translation parts.

Rules for composing and inverting affine transformations

Let $\mathbf{p}, \mathbf{q}$ be any two vectors and let $\mathbf{A}, \mathbf{B}$ be any two invertible matrices.

Rule 4: composition rule

$$(t[\mathbf{p}] \circ \lambda[\mathbf{A}]) \circ (t[\mathbf{q}] \circ \lambda[\mathbf{B}]) = t[\mathbf{p} + \mathbf{Aq}] \circ \lambda[\mathbf{AB}].$$

Rule 5: inverse rule

$$(t[\mathbf{p}] \circ \lambda[\mathbf{A}])^{-1} = t[-\mathbf{A}^{-1}\mathbf{p}] \circ \lambda[\mathbf{A}^{-1}].$$

Proof of the composition rule

$$\begin{aligned}
(t[\mathbf{p}]\,\lambda[\mathbf{A}])\,(t[\mathbf{q}]\,\lambda[\mathbf{B}]) &= t[\mathbf{p}]\,(\lambda[\mathbf{A}]\,t[\mathbf{q}])\,\lambda[\mathbf{B}] \\
&= t[\mathbf{p}]\,(t[\mathbf{Aq}]\,\lambda[\mathbf{A}])\,\lambda[\mathbf{B}] \quad \text{(by Rule 3)} \\
&= (t[\mathbf{p}]\,t[\mathbf{Aq}])(\lambda[\mathbf{A}]\,\lambda[\mathbf{B}]) \\
&= t[\mathbf{p} + \mathbf{Aq}]\,\lambda[\mathbf{AB}] \quad \text{(by Rules 1 and 2)} \quad\blacksquare
\end{aligned}$$

From now on, we shall usually omit the little circle in contexts where it is clear that functions are being composed. Notice that this proof is essentially the same as the solution to Exercise 3.1.

Proof of the inverse rule

$$\begin{aligned}
(t[\mathbf{p}]\,\lambda[\mathbf{A}])^{-1} &= (\lambda[\mathbf{A}])^{-1}\,(t[\mathbf{p}])^{-1} \quad \text{(a property of inverse composites)} \\
&= \lambda[\mathbf{A}^{-1}]\,t[-\mathbf{p}] \\
&= t[-\mathbf{A}^{-1}\mathbf{p}]\,\lambda[\mathbf{A}^{-1}] \quad \text{(by Rule 3)} \quad\blacksquare
\end{aligned}$$

You should recall this property from your previous studies.

Exercise 3.2

Let $f = t[\mathbf{p}]\,\lambda[\mathbf{A}]$, where $\mathbf{p} = (1, -1)$ and

$$\mathbf{A} = \begin{bmatrix} 0 & -1 \\ 1 & 0 \end{bmatrix}.$$

Find f^{-1}, f^2, f^3 and f^4 (where, for example, by f^2 we mean $f \circ f$).

Let us now look at the effect of applying an affine transformation.

Exercise 3.3

Using a piece of squared paper, choose a suitable origin, axes and unit length and draw the square whose vertices are at $(0,0)$, $(1,0)$, $(1,1)$ and $(0,1)$. Then draw the shape to which this square is mapped by the affine transformation

$$\mathbf{x} \mapsto \begin{bmatrix} 1.5 & 0.6 \\ 0.5 & 2.0 \end{bmatrix} \mathbf{x} + \begin{bmatrix} 0.4 \\ -0.2 \end{bmatrix}.$$

What type of shape is this?

> This square is known as the *unit square*.

All affine transformations transform squares into parallelograms. In fact, they always transform straight lines into straight lines, though these new lines may be oriented differently and may have different lengths from the old ones, as Exercise 3.3 illustrated. Furthermore, they always map parallel lines to parallel lines. In summary, although affine transformations preserve linearity and parallelism, they do not in general preserve angles or lengths. Furthermore, it can be shown (and is intuitively obvious) that if a transformation is affine for one particular choice of coordinate system then it will remain affine for any other coordinate system (though the matrix representation will usually be different for each system).

> The fact that they preserve linearity is a direct consequence of their being a composite of a linear transformation and a translation, both of which preserve linearity.

3.3 Isometries

The affine transformation in Exercise 3.3 changes angles and lengths: the unit square changes into a parallelogram whose sides are not of unit length and whose angles are not right angles. It is, of course, perfectly possible to have an affine transformation that does not change angles or lengths. For example, the transformation

$$\mathbf{x} \mapsto \begin{bmatrix} 0 & -1 \\ 1 & 0 \end{bmatrix} \mathbf{x} + \begin{bmatrix} 0.5 \\ 0.5 \end{bmatrix}$$

is of this type, as it is the composite of a rotation by $\pi/2$ and a translation. In fact, once we have specified that lengths do not change, we do not also have to make a specification about angles; for, as will shortly be shown, the constancy of angles follows from the constancy of lengths. We therefore make the following formal definition.

Definition 3.2 Isometry

An **isometry** of $\mathbb{R}^2$ (or of $\mathbb{R}^n$) is a one–one mapping of $\mathbb{R}^2$ (or of $\mathbb{R}^n$) onto itself which preserves the distance between any pair of points.

> Another way to state this is to say that an isometry preserves the lengths of line segments. This means the same as preserving the distances between pairs of points, since the length of a line segment is the distance between its two end-points.

To use this definition in practice, we have to be able to calculate the distance between any pair of points. Recall the formula for the distance between two points in the plane.

> **Definition 3.3 Distance**
>
> The **distance** between two points $A = (a_1, a_2)$ and $B = (b_1, b_2)$ in the plane is
> $$d(A, B) = \sqrt{(a_1 - b_1)^2 + (a_2 - b_2)^2}.$$

This is the same as the distance from the origin to the point $(a_1 - b_1, a_2 - b_2)$, which we call the *length* of the vector $(a_1 - b_1, a_2 - b_2)$. In general, the **length** of the vector $\mathbf{p} = (a, b)$ is $\sqrt{a^2 + b^2}$, and is denoted by $\|\mathbf{p}\|$. Thus, we may now define an isometry as a mapping f from the plane to itself such that

$$\|f(\mathbf{p}) - f(\mathbf{q})\| = \|\mathbf{p} - \mathbf{q}\|, \qquad \text{for all vectors } \mathbf{p} \text{ and } \mathbf{q}. \tag{3.2}$$

If $\|\mathbf{p}\| = 1$ then $\mathbf{p}$ is often referred to as a *unit vector*.

Exercise 3.4

For each function $f : \mathbb{R}^2 \to \mathbb{R}^2$ specified below, explain why it is or is not an isometry.

(a) $f : (x, y) \mapsto (x^2, y^2)$
(b) $f : (x, y) \mapsto (x + 1, y + 1)$
(c) $f : (x, y) \mapsto (x + y, x - y)$
(d) $f : (x, y) \mapsto (y, x)$
(e) $f : (x, y) \mapsto \left(\tfrac{1}{2}x - \tfrac{1}{2}y\sqrt{3}, \tfrac{1}{2}x\sqrt{3} + \tfrac{1}{2}y\right)$

Hint When looking for counter-examples to the distance-preserving property, try simple cases for $\mathbf{p}$ and $\mathbf{q}$ first.

Exercise 3.5

Which of the functions in Exercise 3.4 are affine transformations? For those that are, give the linear and the translation parts.

Exercise 3.4 shows that checking specific examples to determine whether they are or are not isometries can be very time-consuming. What is needed is a result of the type: 'All plane isometries are of the form ...'. Such a result is the subject of the next subsection.

3.4 A characterization of isometries

Consider the functions in Exercise 3.4. We have seen that all those that are isometries are affine transformations, and it seems reasonable to conjecture that this may be true in general. Also, we have seen that it is particularly easy to show that the function in part (b) (which is just a translation) is an isometry, and this suggests that *any* translation is an isometry. This second conjecture is in fact quite easy to prove, and you are now asked to do this.

Exercise 3.6

Show that, for any vector $\mathbf{p}$, the translation $t[\mathbf{p}]$ is an isometry.

It is also true that any isometry is an affine transformation, as we shall prove at the end of this subsection. The proof of this requires the notion of the *dot* (or *scalar*) *product* of two vectors.

This is another concept that you have probably met in your previous studies.

> **Definition 3.4 Dot product**
>
> The **dot product** of the vectors $\mathbf{p} = (a, b)$ and $\mathbf{q} = (c, d)$ is the real number
>
> $$\mathbf{p} \cdot \mathbf{q} = ac + bd.$$
>
> That is to say, it is the product of the two x-coordinates plus the product of the two y-coordinates.

The corresponding definition in three or more dimensions is very similar: the dot product is just the sum of the products of corresponding pairs of coordinates.

Exercise 3.7

Show that, for any three vectors $\mathbf{p}$, $\mathbf{q}$ and $\mathbf{r}$:

(a) $\mathbf{p} \cdot \mathbf{q} = \mathbf{q} \cdot \mathbf{p}$;

(b) $(\mathbf{p} + \mathbf{q}) \cdot \mathbf{r} = \mathbf{p} \cdot \mathbf{r} + \mathbf{q} \cdot \mathbf{r}$;

(c) $\mathbf{p} \cdot (\mathbf{q} + \mathbf{r}) = \mathbf{p} \cdot \mathbf{q} + \mathbf{p} \cdot \mathbf{r}$.

As you may recall from previous mathematics work, both *length* and *angle* can be described in terms of dot products. If $\mathbf{p}$ is any vector, then the *length* of $\mathbf{p}$ is given by

$$\|\mathbf{p}\| = \sqrt{\mathbf{p} \cdot \mathbf{p}}, \tag{3.3}$$

and if $\mathbf{p}$ and $\mathbf{q}$ are any non-zero vectors, then the *angle* between $\mathbf{p}$ and $\mathbf{q}$ (that is, the angle between the lines from the origin to the corresponding points P and Q) can be characterized by its cosine, using the equation

$$\cos \theta = \frac{\mathbf{p} \cdot \mathbf{q}}{\|\mathbf{p}\| \, \|\mathbf{q}\|}. \tag{3.4}$$

Note that, for unit vectors $\mathbf{p}$, Equation 3.3 tells us that $\mathbf{p} \cdot \mathbf{p} = 1$.

Now, Equation 3.2 above characterizes isometries in terms of lengths, while Equation 3.3 expresses lengths in terms of dot products. Putting these together, we can characterize an isometry as a mapping f from the plane to itself such that

$$\sqrt{\big(f(\mathbf{p}) - f(\mathbf{q})\big) \cdot \big(f(\mathbf{p}) - f(\mathbf{q})\big)} = \sqrt{(\mathbf{p} - \mathbf{q}) \cdot (\mathbf{p} - \mathbf{q})};$$

or equivalently (since we are taking positive square roots) such that

$$\big(f(\mathbf{p}) - f(\mathbf{q})\big) \cdot \big(f(\mathbf{p}) - f(\mathbf{q})\big) = (\mathbf{p} - \mathbf{q}) \cdot (\mathbf{p} - \mathbf{q}). \tag{3.5}$$

Note that it is conventional to specify the angle between two vectors as an angle in the half-open interval $[0, \pi[$. Note also, from Equation 3.4, that $\theta = \pi/2$ if and only if $\mathbf{p} \cdot \mathbf{q} = 0$, i.e. two vectors are at right-angles if and only if their dot product is zero.

In order to make further progress, we now temporarily confine our attention to isometries that fix the origin, i.e. those for which $f(\mathbf{0}) = \mathbf{0}$.

Exercise 3.8

Show that if f is an isometry of $\mathbb{R}^2$ that fixes the origin, then it preserves dot products: that is,

$$f(\mathbf{p}) \cdot f(\mathbf{q}) = \mathbf{p} \cdot \mathbf{q},$$

for any two vectors $\mathbf{p}$, $\mathbf{q}$ in $\mathbb{R}^2$. Use Equation 3.4 to deduce that the magnitudes of angles are also preserved by such isometries.

Hint For any vector $\mathbf{p}$, we can write $\mathbf{p} = \mathbf{p} - \mathbf{0}$, where $\mathbf{0} = (0, 0)$ is the zero vector representing the origin.

We are now ready to prove a result which is an important stepping stone on the way to characterizing isometries.

> *Lemma 3.1*
>
> Any isometry that fixes the origin is a linear transformation.

Proof

Let f be an isometry, and let $f(0,0) = \mathbf{p}$. Consider the function

$$\lambda(\mathbf{x}) = f(\mathbf{x}) - \mathbf{p}, \quad \text{for all } \mathbf{x} \text{ in } \mathbb{R}^2. \tag{3.7}$$

It is the composite of two isometries, namely f and translation by $-\mathbf{p}$. Thus, by Theorem 3.1, λ is an isometry. But, substituting $(0,0)$ for $\mathbf{x}$ in Equation 3.7 gives

Recall, from Exercise 3.6, that all translations are isometries.

$$\lambda(0,0) = f(0,0) - \mathbf{p} = \mathbf{p} - \mathbf{p} = (0,0),$$

so that λ fixes the origin. Thus, by Lemma 3.1, $\lambda = \lambda[\mathbf{A}]$ for some matrix $\mathbf{A}$. Furthermore, from the observation immediately following Lemma 3.1, the matrix $\mathbf{A}$ is orthogonal.

Now, Equation 3.7 can be rewritten as

$$f(\mathbf{x}) = \lambda[\mathbf{A}](\mathbf{x}) + \mathbf{p} = t(\mathbf{p}) \circ \lambda[\mathbf{A}](\mathbf{x}),$$

and so f is an isometry whose linear part is $\lambda[\mathbf{A}]$, for some orthogonal matrix $\mathbf{A}$, and whose translation part is $t[\mathbf{p}]$, where $\mathbf{p} = f(0,0)$.

Conversely, suppose that $f = t[\mathbf{p}] \circ \lambda[\mathbf{A}]$, where $\mathbf{A}$ is the orthogonal matrix

$$\mathbf{A} = \begin{bmatrix} a & c \\ b & d \end{bmatrix}.$$

Let $\mathbf{q} = (p,q)$ and $\mathbf{r} = (r,s)$ be any two points in $\mathbb{R}^2$. Then

$$\lambda[\mathbf{A}](\mathbf{q}) - \lambda[\mathbf{A}](\mathbf{r}) = \lambda[\mathbf{A}](\mathbf{q} - \mathbf{r}) \quad \text{(by the linearity of } \lambda[\mathbf{A}]),$$

and so

$$\begin{aligned}
\|\lambda[\mathbf{A}](\mathbf{q}) - \lambda[\mathbf{A}](\mathbf{r})\|^2 &= \|\lambda[\mathbf{A}](p-r, q-s)\|^2 \\
&= \|(a(p-r) + c(q-s), b(p-r) + d(q-s))\|^2 \\
&= \bigl(a(p-r) + c(q-s)\bigr)^2 + \bigl(b(p-r) + d(q-s)\bigr)^2 \\
&= a^2(p-r)^2 + 2ac(p-r)(q-s) + c^2(q-s)^2 \\
&\quad + b^2(p-r)^2 + 2bd(p-r)(q-s) + d^2(q-s)^2 \\
&= \bigl(a^2 + b^2\bigr)(p-r)^2 + \bigl(c^2 + d^2\bigr)(q-s)^2 \\
&\quad + 2(ac + bd)(p-r)(q-s) \\
&= (p-r)^2 + (q-s)^2 \quad \text{(by the orthogonality of } \mathbf{A}) \\
&= \|\mathbf{q} - \mathbf{r}\|^2.
\end{aligned}$$

See Equations 3.6.

Therefore

$$\|\lambda[\mathbf{A}](\mathbf{q}) - \lambda[\mathbf{A}](\mathbf{r})\| = \|\mathbf{q} - \mathbf{r}\| \quad \text{(taking positive square roots)}.$$

Thus, $\lambda[\mathbf{A}]$ is an isometry. But we know, from Exercise 3.6, that $t[\mathbf{p}]$ is an isometry. Hence, by Theorem 3.1, $f = t[\mathbf{p}] \circ \lambda[\mathbf{A}]$ is an isometry. ∎

Exercise 3.10

Use Theorems 3.1 and 3.2, together with the rule for composing affine transformations, to show that:

(a) the product of two orthogonal matrices is orthogonal;

(b) the inverse of an orthogonal matrix is orthogonal.

In the next section we return to the subject of tilings. In particular, we study tilings using polygonal tiles all of which are congruent (that is, for any two tiles, there is an isometry which maps one onto the other). We shall, however, drop the edge-to-edge requirement which we considered in Section 2.

4 TILINGS USING CONGRUENT POLYGONS

4.1 Monohedral tilings

The word *congruence* has slightly different usages in different areas of mathematics. In particular, its usage in the present geometric context differs somewhat from its usage in number theory. In order to avoid any confusion, we make the following definition.

> *Definition 4.1 Congruence*
>
> Two shapes in the plane are **congruent** if there is an isometry which maps one exactly onto the other.

In many contexts, we expect all the individual tiles of a tiling to be congruent. A set of bathroom tiles, for example, usually consists of rectangles all of the same size (and hence congruent). A tiling constructed of congruent tiles is said to be *monohedral*.

> *Definition 4.2 Monohedral tiling and template*
>
> A tiling $\mathcal{T}$ is **monohedral** if all the tiles of $\mathcal{T}$ are congruent to one single shape, the **template** for $\mathcal{T}$.

This definition does not require the tiles to be polygons, but we shall consider only polygonal tiles in this section.

Exercise 4.1

Which of the tilings depicted in Figure 1.1 are monohedral?

Remember, from Exercise 1.3, that only (a)–(g) and (l) in Figure 1.1 depict tilings.

You may be surprised to learn that there are infinitely many monohedral tilings, despite the fact that restricting the tiles to be congruent to a single shape seems such a drastic restriction.

For a start, we can change scales in the x- and y-directions and convert our square tiling $\mathcal{R}_4$ into a rectangular one. Such a scaling also changes $\mathcal{R}_3$ and $\mathcal{R}_6$ into tilings that are still monohedral, but in each case the template is not a *regular* polygon.

$\mathcal{R}_3$, $\mathcal{R}_4$ and $\mathcal{R}_6$ are shown in Figure 2.6 and on Side 1 of Tiling Card 1.

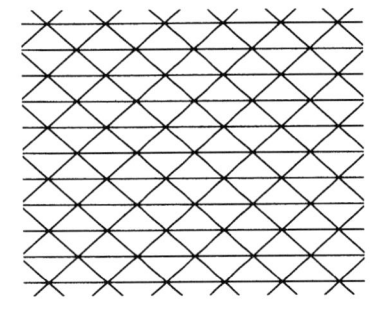

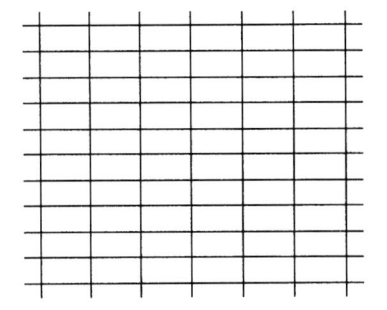

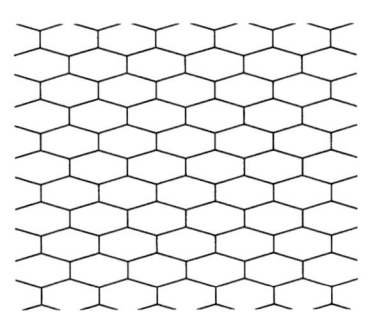

Figure 4.1 Examples of tilings obtained by scaling $\mathcal{R}_3$, $\mathcal{R}_4$ and $\mathcal{R}_6$.

We can even skew the axes as well (or instead), so that the squares of $\mathcal{R}_4$ become parallelograms. Again, the effect on $\mathcal{R}_3$ and $\mathcal{R}_6$ is to produce monohedral tilings, and again in each case the template is not a regular polygon.

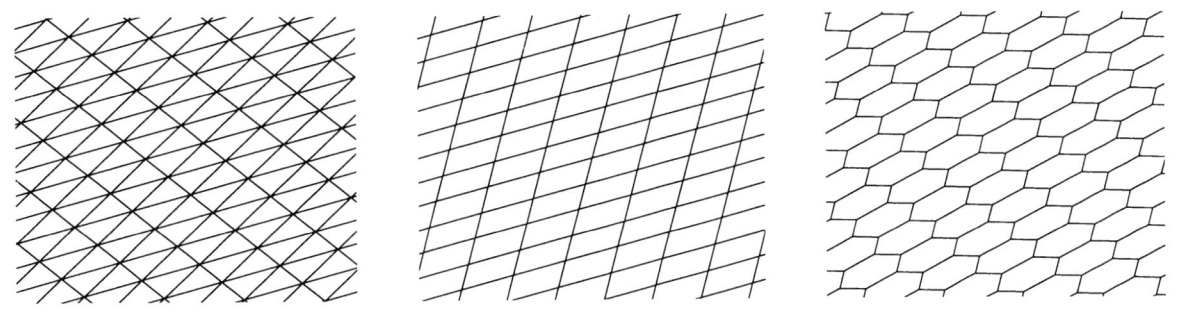

Figure 4.2 Examples of tilings obtained by scaling and skewing $\mathcal{R}_3$, $\mathcal{R}_4$ and $\mathcal{R}_6$.

These scalings and skewings are obtained by applying affine transformations to the plane on which the original tiling is drawn.

You may wonder whether *every* possible shape of triangle, quadrilateral and hexagon can give rise to a tiling which can be obtained from a regular tiling in this way. In order to answer this, we have to ask whether every triangle, quadrilateral or hexagon can be produced from a regular one by applying an affine transformation.

The answer as far as triangles are concerned is *yes*. You may recall from previous work that an important theorem of affine geometry states that *any triangle can be mapped exactly onto any other triangle* by means of a suitable affine transformation. In particular, an equilateral triangle can be mapped onto any triangle whatsoever. Furthermore, we can deduce from this theorem that any parallelogram can be mapped exactly onto any other parallelogram by means of a suitable affine transformation. And, in particular, a square can be mapped onto any parallelogram whatsoever.

We postpone answering our question in the case of quadrilaterals in general and in the case of hexagons until later in this subsection, though you may like to speculate as to the likely answer.

This theorem is sometimes called the Fundamental Theorem of Affine Geometry. This is the name that we shall use for a strengthened form of the theorem, from which the above statements on triangles and parallelograms follow as a corollary.

The proof of this theorem and of its corollary is *optional*, and appears in the Appendix.

Theorem 4.1 Fundamental theorem of affine geometry

Let P, Q, R be any three non-collinear points in $\mathbb{R}^2$, and let U, V, W be any three other such points. Then there is exactly one affine transformation that maps P to U, Q to V and R to W.

Non-collinear means that there is no straight line passing through all three points.

Corollary 4.1

Any triangle can be mapped exactly onto any other triangle, and any parallelogram can be mapped exactly onto any other parallelogram, by a suitable affine transformation.

There are actually six 'suitable' affine transformations in the case of triangles, and eight in the case of parallelograms. Can you see why?

We now have to be a little careful. We have just observed that an equilateral triangle can be mapped onto any triangular shape. However, this does not tell us that the result of applying an affine transformation to the whole tiling $\mathcal{R}_3$ produces a monohedral tiling! What if the affine transformation maps two congruent triangles to two non-congruent triangles?

You may wish to brush this possibility contemptuously aside. But before doing so, consider the tiling depicted in Figure 4.3(a) (whose pattern is commonly found on wooden tiled floors). Each tile measures 2 units by 1 unit. Suppose now that we apply an affine transformation whose linear part is given by the matrix

$$\begin{bmatrix} 2 & 0 \\ 0 & 1 \end{bmatrix}$$

(the translation part is irrelevant). This scales up in the x-direction by a factor of 2, so that the tiles which had been laid with their long sides parallel to the x-axis now measure 4 by 1, while those which had been laid with their long sides parallel to the y-axis now measure 2 by 2, as Figure 4.3(b) illustrates. The tiling is no longer monohedral!

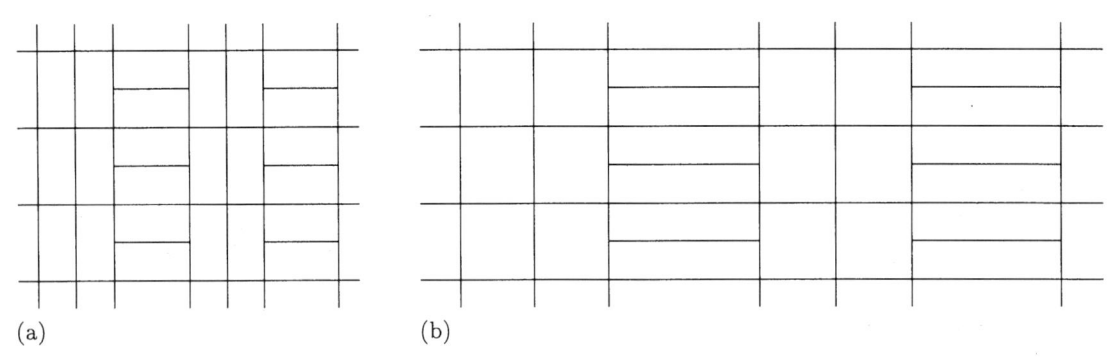

(a)   (b)

Figure 4.3

Despite this warning, it is in fact true that applying an affine transformation to any of the tilings $\mathcal{R}_3$, $\mathcal{R}_4$ or $\mathcal{R}_6$ does produce a monohedral tiling. This means that some monohedral tilings *are* always mapped to monohedral tilings by affine transformations, while others (such as the one depicted in Figure 4.3(a)) can be mapped to *non*-monohedral tilings by affine transformations.

In the cases of $\mathcal{R}_4$ and $\mathcal{R}_6$, the reason why applying an affine transformation always produces another monohedral tiling is that every tile of the tiling has the same *orientation*. That is, every tile is 'the same way up', so that any tile can be mapped to any other using a translation. After the application of an affine transformation, this 'all the same way up' property still holds, and, since a translation is always an isometry, this means that after the affine transformation the tiling is still monohedral.

Remember that isometries preserve straight lines, lengths and angles.

In the case of $\mathcal{R}_3$ this simple reasoning is not sufficient, as the triangles have two orientations, or two ways up. Every triangle has one edge drawn horizontally on the page. Regarding this edge as the base and the opposite vertex as the apex, half the triangles have the apex above the base and half have the apex below the base. (More intuitively, half point up and half point down.) Now, the two orientations can be interchanged by a rotation through π. This is important, because if two congruent shapes are placed in the plane with their orientations either the same or differing by π, then any affine transformation of the plane leaves them congruent. This result, which means that any affine transformation of $\mathcal{R}_3$ must always produce a monohedral tiling, is worth formally recording, through it is not quite interesting enough to be called a theorem.

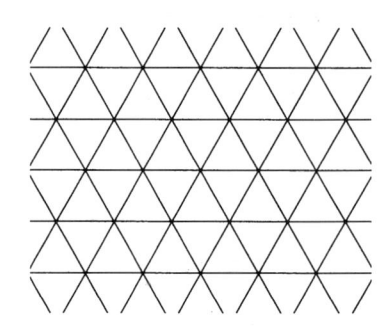

Figure 4.4 $\mathcal{R}_3$.

> **Lemma 4.1 Affine congruency lemma**
>
> Let S and T be two plane figures, such that S can be mapped exactly onto T by means of an isometry whose linear part is either the identity or rotation through π. Let f be any affine transformation of the plane. Then $f(S)$ is congruent to $f(T)$.

Notice that the fact that this lemma holds for isometries whose linear part is the identity (as well as for isometries whose linear part is rotation through π) provides us with a formal proof that any affine transformation of $\mathcal{R}_4$ or $\mathcal{R}_6$ must always produce a monohedral tiling.

Proof

Let g be the isometry which maps S to T. Then $f(S)$ can be mapped to $f(T)$ by applying the inverse of f (to get from $f(S)$ to S), followed by g (to get to T), followed by f (to get to $f(T)$). That is to say, $f(T) = h(f(S))$, where

$$h = fgf^{-1}.$$

Thus, all we have to do is to show that h is an isometry.

By the work of the previous section, we can decompose f and g into linear and translation parts:

$$f = t[\mathbf{p}]\, \lambda[\mathbf{A}], \quad g = t[\mathbf{q}]\, \lambda[\pm\mathbf{I}]$$

(where $\mathbf{I}$ denotes the identity matrix). Now the composition rule for affine transformations tells us that, when composing affine transformations, we can find the matrix for the linear part of the composite simply by multiplying the matrices for the individual linear parts. In this case, therefore, the linear part of h has the matrix $\mathbf{A}(\pm\mathbf{I})\mathbf{A}^{-1}$, which is simply $\pm\mathbf{I}$. Therefore, h is indeed an isometry, and so $f(S)$ is congruent to $f(T)$. ∎

A rotation through π about the origin simply reverses the sign of each coordinate of any point, and so is represented by the matrix $-\mathbf{I}$.

Exercise 4.2

The tiles in Figure 4.3(a) are in two orientations. By what angle must a tile in one orientation be rotated in order to map it to a tile in the other?

The Affine Congruency Lemma tells us that any affine transformation of $\mathcal{R}_3$, $\mathcal{R}_4$ or $\mathcal{R}_6$ will always produce a monohedral tiling. In the case of $\mathcal{R}_3$, the lemma, combined with the fact that an equilateral triangle can be mapped to any triangle by an affine transformation (which is a consequence of the Fundamental Theorem of Affine Geometry), also shows that, for *any* triangular template T, we can find an affine transformation that maps $\mathcal{R}_3$ to a monohedral tiling using T. However, the corresponding results are *not* true for $\mathcal{R}_4$ for $\mathcal{R}_6$, though, as we shall see, we can draw some conclusions about monohedral tilings involving quadrilaterals or hexagons.

A property of affine transformations is that they map parallel lines to parallel lines. Another property is that two parallel line segments of equal length are mapped to another two line segments of equal length (though not usually of the same length as the first two). Now squares and regular hexagons have opposite sides equal in length and parallel. Thus, under affine transformations, they must map to figures with opposite sides equal in length and parallel. In the case of quadrilaterals, as you will be aware, such figures are called *parallelograms*. There is no special name for hexagons of this type, so we invent one.

You should have come across this second property of affine transformations before. We do not prove it here, though it is not difficult to prove; you may like to try to prove it for yourself.

> **Definition 4.3 Parallelohexagon**
>
> A **parallelohexagon** is a hexagon with opposite sides parallel and equal in length.

Thus, any figure obtained by applying an affine transformation to a square
or regular hexagon must be a parallelogram or a parallelohexagon. But can
every parallelogram and parallelohexagon be obtained in this way? This
time the answer is *yes* for parallelograms, but *no* for parallelohexagons.

One way to see why not every parallelohexagon can be obtained by an affine
transformation of a regular hexagon is to draw the diagonals of a regular
hexagon, as in Figure 4.5. This splits the hexagon into six *equilateral*
triangles, three with one orientation and three with the opposite orientation
(i.e. rotated through π with respect to the first three). Thus (by the Affine
Congruency Lemma), when we apply an affine transformation we obtain a
parallelohexagon whose diagonals split it into six *congruent* triangles, as
shown in Figure 4.6. However, if you draw an *arbitrary* parallelohexagon
(which you can do by drawing three successive sides, then drawing the other
three equal in length and parallel to the first three), and join its diagonals,
the six triangles so obtained are in general *not* congruent, as Figure 4.7
illustrates. Thus an arbitrary parallelohexagon is not usually the image of a
regular hexagon under an affine transformation.

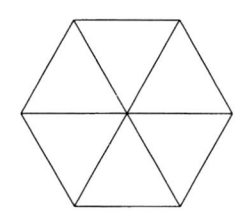

Figure 4.5

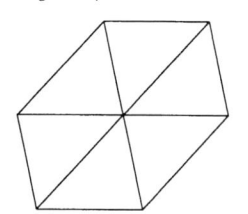

Figure 4.6

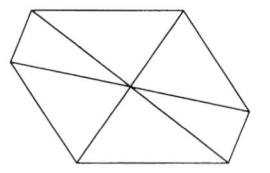

Figure 4.7

Nevertheless, any parallelohexagon (even one not obtained from an affine
transformation of a regular hexagon) *can* be used as the template for a
monohedral, edge-to-edge tiling! We prove this now, as part of a more
general theorem, which also contains a proof of why every parallelogram can
be obtained by applying an affine transformation to a square.

Theorem 4.2 Monohedral tiling theorem

Any triangle, quadrilateral or parallelohexagon can be the template for
a monohedral tiling. Moreover, if the template is a triangle or a
parallelogram, then such a tiling can be obtained from $\mathcal{R}_3$ or from $\mathcal{R}_4$,
respectively, by applying an appropriate affine transformation of the
plane.

Proof

The 'moreover' part of the theorem is the easiest part to prove, so we shall
do it first. Then we shall prove the result for parallelohexagons, and finally
we shall mop up those quadrilaterals that are not parallelograms.

Triangles and parallelograms

Let T be any triangle or parallelogram, and let $\mathcal{T}$ be the tiling $\mathcal{R}_3$ if T is a
triangle or $\mathcal{R}_4$ if T is a parallelogram. By the Fundamental Theorem of
Affine Geometry, there is an affine transformation f of the plane that maps
one tile of $\mathcal{T}$ exactly onto T. But *either* every tile of $\mathcal{T}$ is of the same
orientation as every other (if $\mathcal{T}$ is $\mathcal{R}_4$) *or* the orientations differ only by
rotations through π (if $\mathcal{T}$ is $\mathcal{R}_3$). Thus, by the Affine Congruency Lemma, all
the tiles of $f(\mathcal{T})$ are congruent; that is to say, $f(\mathcal{T})$ is a monohedral tiling
whose template is T, and which is obtained by applying an affine
transformation to $\mathcal{R}_3$ or $\mathcal{R}_4$, as required.

Parallelohexagons

Let P be any parallelohexagon. If we label the sides of such a figure s_1 to s_6,
then we can surround it by six congruent figures, P_1 to P_6, as shown in
Figure 4.8. Side s_1 of P is the same as side s_4 of P_1; side s_2 of P is the same
as side s_5 of P_2; etc. Continuing in this way, we can tile the whole plane,
because at each vertex we get one copy of each of the three different internal
angles of P, and these sum to 2π, as required.

Notice that the tiling we construct is edge-to-edge.

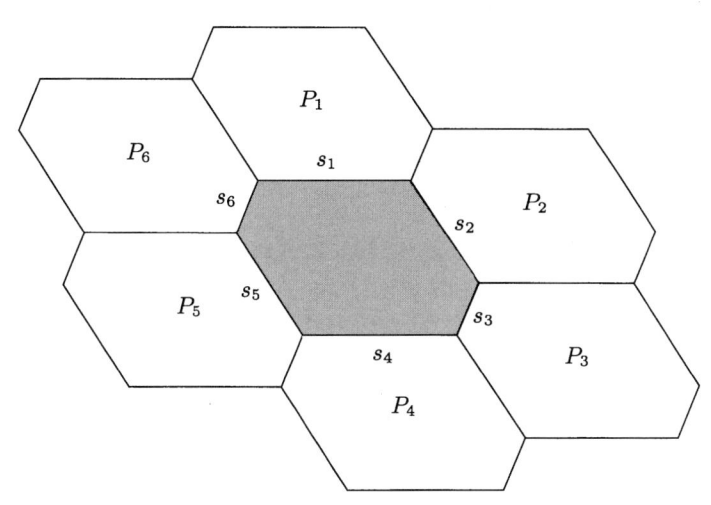

Figure 4.8

General quadrilaterals

Let Q be any quadrilateral, with vertices a, b, c, d and internal angles $\alpha, \beta, \gamma, \delta$, as shown in Figure 4.9(a). Draw another copy of Q, sharing the edge ab with Q, by *rotating* Q by a half-turn about the midpoint of ab, as shown in Figure 4.9(b). The vertices of this copy are a_1, b_1, c_1, d_1 (where a_1 coincides with b and b_1 coincides with a).

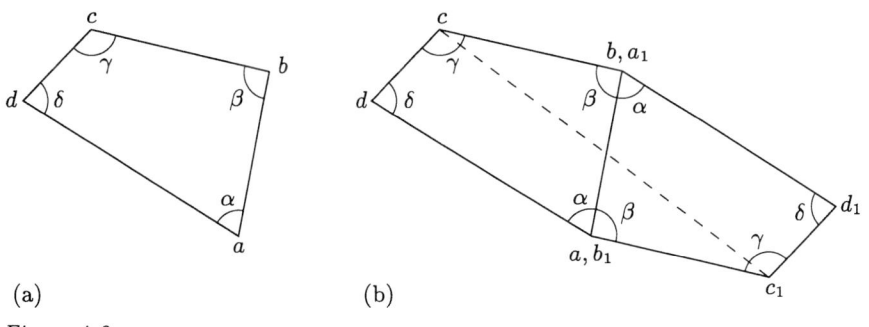

(a)   (b)

Figure 4.9

The opposite angles α show that ad is parallel to $a_1 d_1$, and the opposite angles β show that bc is parallel to $b_1 c_1$. It is also true that cd is parallel to $c_1 d_1$. There are various ways to show this; possibly the easiest is to join c and c_1. The angles $cc_1 b_1$ and $c_1 cb$ are equal (as bc and $b_1 c_1$ are parallel); hence the angles $cc_1 d_1$ and $c_1 cd$ are equal; thus cd and $c_1 d_1$ are parallel. ∎

Such pairs of angles are sometimes referred to as *alternate angles*.

It follows that the whole figure is a parallelohexagon. But we have just seen that any parallelohexagon can be a template for a monohedral tiling. Dividing each parallelohexagon of such a tiling in two by the appropriate diagonal gives a required tiling by tiles congruent to Q.

Exercise 4.3

Draw a tiling of the plane (i.e. draw enough tiles to satisfy yourself that you understand the pattern) using the template shown in Figure 4.10. (You might find it easiest to do this by drawing your tiling on a thin piece of paper, so that you can place this figure underneath it and trace its shape through the paper.)

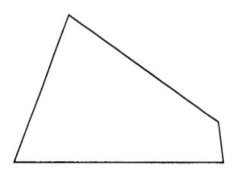

Figure 4.10

You might be forgiven for thinking that we have now covered all possible monohedral tilings, or at least all those that are polygonal. But this is very far from being so. Even restricting ourselves to rectangular tiles, we have examples such as those in Figure 4.11.

Remember that we do not in this section require the tilings to be edge-to-edge.

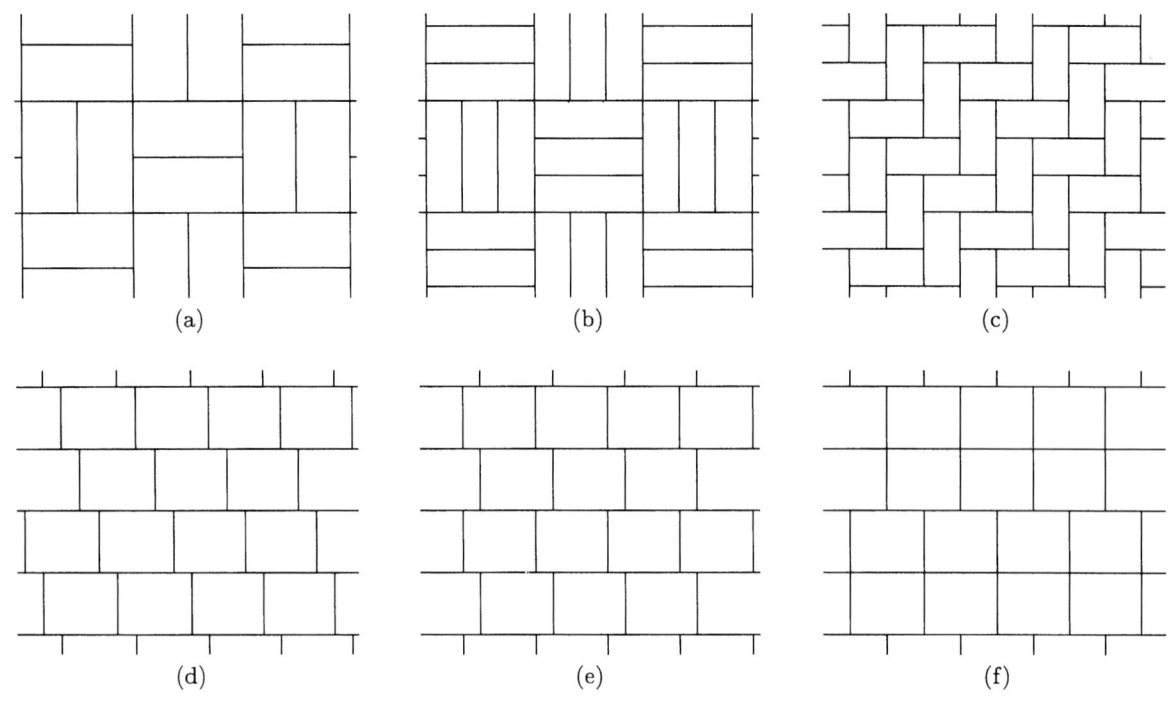

Figure 4.11

The first two of these examples are actually members of an *infinite* sequence of monohedral rectangular tilings! Starting with $\mathcal{R}_4$, we can define a tiling $\mathcal{T}_n$ for any value of n (≥ 2) by slicing the squares of $\mathcal{R}_4$ into n rectangles, alternate squares being sliced horizontally and vertically. Thus the first tiling in Figure 4.11 is $\mathcal{T}_2$, while the second is $\mathcal{T}_3$. The third example is quite often seen on floors and on paved pedestrian precincts. The fourth and fifth are sometimes observed in brickwork. The sixth is included for the sake of further variety.

Many fascinating monohedral tilings are known, such as those depicted on Side 2 of Tiling Card 2 in the *Geometry Envelope*. There seems to be no hope of classifying all of these!

If you are interested in reading further on the subject of classifying monohedral tilings, you should consult *Tilings and Patterns*, by B. Grünbaum and G.C. Shephard (see the Bibliography in the *Course Guide*).

4.2 The Laves tilings

When we considered the Archimedean tilings, we noted that, in each tiling, all the vertices are of the same type. That is to say, each vertex has the same arrangement of incident tiles. It is interesting to ask what happens when we exchange the roles of vertices and tiles, and ask that each *tile* has the same arrangement of incident *vertices*. We make the following definition.

Definition 4.4 Tile type

Let T be any tile in a tiling $\mathcal{T}$. The **tile type** of T is given by listing the degrees of the vertices incident with T, starting with any one of them and proceeding in either clockwise or anticlockwise order. This list is conventionally placed in square brackets (to distinguish it from the symbol for vertex type).

For example, a tile type [3, 3, 3, 3, 3, 3] (characteristic of the regular tiling $\mathcal{R}_6$) records the fact that there are six vertices incident with any tile, all of degree 3 (see Figure 4.12).

As in the case of vertex types, we regard two tile types as being the same if we can make them identical by starting at an appropriate vertex and then proceeding in an appropriate direction.

Figure 4.12

Definition 4.5 Equality of tile types

Two tile types $[a_1, a_2, \ldots, a_n]$ and $[b_1, b_2, \ldots, b_n]$ are the **same** if, for some i ($0 \leq i \leq n-1$), either $a_j = b_{i+j}$ ($j = 1, \ldots, n$) or $a_j = b_{i-j}$ ($j = 1, \ldots, n$), where we interpret the subscripts modulo n.

Furthermore, analogous to the concept of vertex-uniformity, we have a concept of *tile-uniformity*.

Definition 4.6 Tile-uniform tiling

A tiling is **tile-uniform** if all its tile types are the same.

Exercise 4.4

Each of the tilings in Figure 4.11 except (b) is tile-uniform. Write down each of the corresponding tile types.

Exercise 4.5

Some of the Archimedean tilings are tile-uniform (as well as being vertex-uniform). Which are they? Compare their tile types with their vertex types.

Remember that the Archimedean tilings are those on Tiling Card 1 in the *Geometry Envelope*.

The result arising from Exercise 4.5 is interesting, as it suggests a connection between the tilings $\mathcal{R}_3$ and $\mathcal{R}_6$: the tile type of $\mathcal{R}_3$ is the same as the vertex type of $\mathcal{R}_6$ (except for a different bracket style), and vice versa. It appears that the tiles of the one tiling may correspond to the vertices of the other in some way.

A graphical way to see this connection is to draw $\mathcal{R}_3$ with thick lines, then to place a dot at the centre of each tile and join with a thin line those dots corresponding to adjacent tiles, as shown in Figure 4.13.

Exercise 4.6

The thick lines in Figure 4.13 represent the tiling $\mathcal{R}_3$. What do the thin lines represent?

Notice that, in Figure 4.13, the vertices of $\mathcal{R}_3$ are at the centres of the hexagons making up the tiling $\mathcal{R}_6$. Thus, starting with $\mathcal{R}_6$, we could perform the same process as we did in drawing Figure 4.13 and produce the tiling $\mathcal{R}_3$.

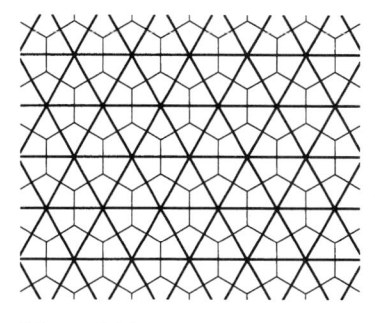

Figure 4.13

We say that two tilings such as $\mathcal{R}_3$ and $\mathcal{R}_6$, where one can be constructed from the other in the above way, are *dual* tilings. The general definition of duality for polygonal edge-to-edge tilings is as follows.

> *Definition 4.7 Dual tilings*
>
> Let S and $\mathcal{T}$ be polygonal edge-to-edge tilings. Then $\mathcal{T}$ is **dual** to S if $\mathcal{T}$ can be constructed from S by placing the vertices of $\mathcal{T}$ at the centres of the tiles of S and joining two vertices of $\mathcal{T}$ (by a straight line) if and only if the corresponding tiles of S are adjacent.

This definition is deliberately imprecise, in that it does not specify what is meant by the 'centre' of a tile. A more rigorous approach to duality is complicated and fiddly, and beyond the scope of this course. This somewhat informal approach is adequate for our purposes.

Exercise 4.7

Which tiling is dual to $\mathcal{R}_4$?

We have seen that the dual of the tiling with vertex type $(3,3,3,3,3,3)$ has tile type $[3,3,3,3,3,3]$, while the dual of the tiling with vertex type $(6,6,6)$ has tile type $[6,6,6]$. Moreover, we have seen that the square tiling is dual to itself and has vertex type $(4,4,4,4)$ and tile type $[4,4,4,4]$. The obvious next question is what happens if we perform the duality construction (placing a dot at the centre of each tile, and then joining dots corresponding to adjacent tiles) on each of the remaining eight Archimedean tilings. The results of performing the duality construction for all eleven Archimedean tilings is shown in Figure 4.14.

Removing the original vertex-uniform tiling in each case reveals the new tile-uniform tiling. The eleven tilings obtained in this way appear on Tiling Card 3 in the *Geometry Envelope*. Note that, just as $(3,3,3,3,6)$ has two mirror-image forms, so does $[3,3,3,3,6]$, as you can check by turning over Overlay 1 for Side 1 of Tiling Card 3 and then trying to superimpose the overlay picture over the card picture for this tiling. Also, just as $(4,6,12)$ can be considered as having two types of *vertex*, one the mirror image of the other, so $[4,6,12]$ has two types of *tile*, one the mirror image of the other. These eleven tilings are known as the **Laves tilings** (pronounced 'lah-veys'), after the early twentieth-century German crystallographer Fritz Laves.

You may wish to convince yourself of the duality between the Archimedean and the Laves tilings, by using Overlays 1 for Tiling Card 3 in conjunction with Tiling Card 1 and by using Overlays 1 for Tiling Card 1 in conjunction with Tiling Card 3.

Exercise 4.8

Which of the Laves tilings are also Archimedean tilings?

Exercise 4.9

Which of the Laves tilings have nets consisting simply of sets of infinite parallel straight lines?

Exercise 4.10

Find the degree of a typical tile in each of the Laves tilings.

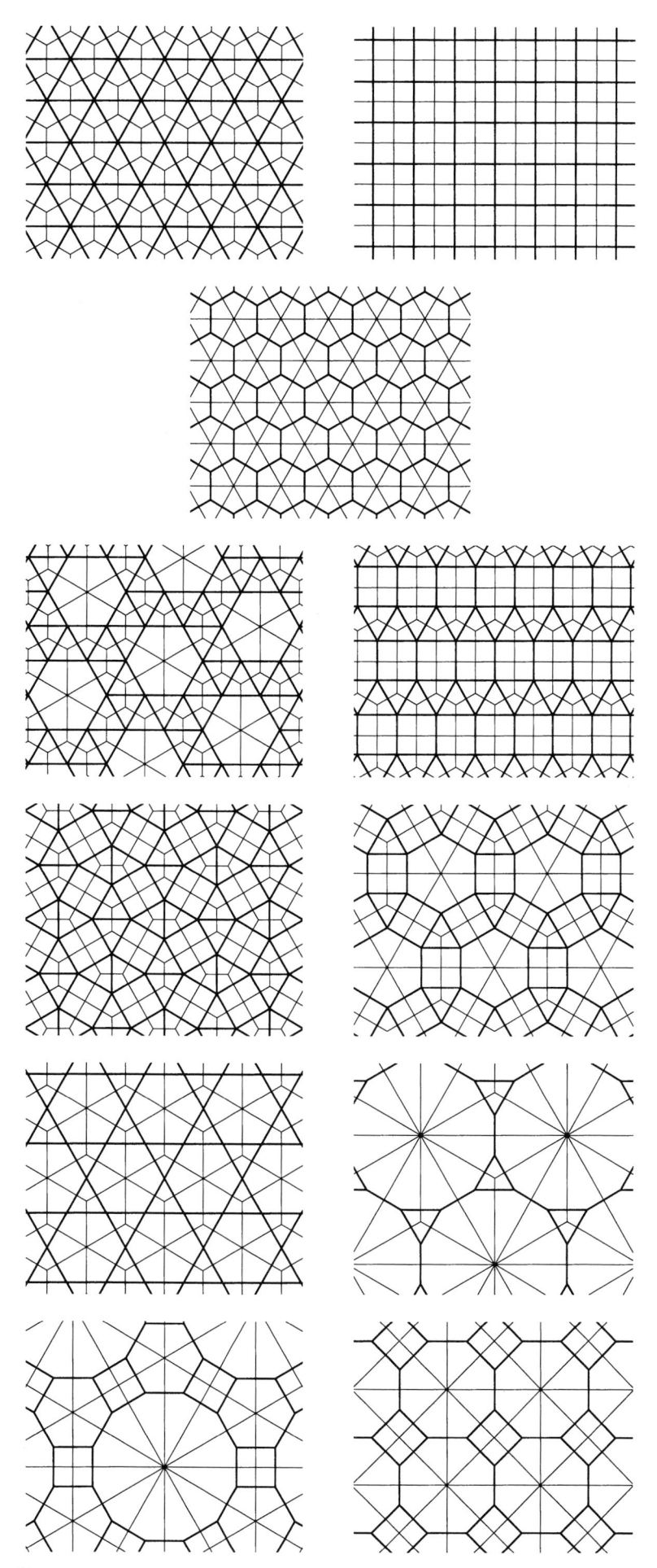

Figure 4.14

5 WORKING WITH ISOMETRIES

In Section 3 we developed an *algebraic* characterization of all plane isometries. Given a coordinate system for the plane and a description of a function from the plane to itself, we can tell from this characterization whether that function is an isometry. We check that the function is affine (i.e. is the composite of a linear part defined by a matrix and a translation part defined by a vector), then check whether the matrix of the linear part is orthogonal.

This is not a great deal of use, however, if you are given some isometries described *geometrically* and are asked to compose them. For example, what is the result of performing first an anticlockwise rotation by $\pi/3$ about the point $(\sqrt{3}, 1)$, and then a reflection in the line $y = x + 3$? Clearly, if we wish to use the rules of Section 3 to compose these, we need to know how to express each of them as the composite $t[\mathbf{p}] \, \lambda[\mathbf{A}]$, where $\mathbf{A}$ is orthogonal. We shall call this form of writing an isometry the **standard form** of an isometry.

In this section, we shall first go through the various geometric properties which a *plane isometry* (i.e. an isometry of the plane) can have, and then develop Rules 1 to 3 of Section 3 into a set of equations relating the various ways of expressing isometries. We have called these equations our **Isometry Toolkit**, and they are printed on the card of that name in the *Geometry Envelope*.

You are advised to have this card with you when you study *Unit IB3* and the other Geometry units.

5.1 A geometric classification of isometries

It is convenient to begin with isometries that fix the origin — those that are defined just by an orthogonal matrix, without any translation.

It is easy to give a neat characterization of 2×2 orthogonal matrices. Suppose the matrix

$$\mathbf{A} = \begin{bmatrix} a & c \\ b & d \end{bmatrix}$$

is orthogonal. Then, because $a^2 + b^2 = 1$, the point (a, b) lies on a circle whose centre is the origin and whose radius is 1. We can therefore find an angle θ such that

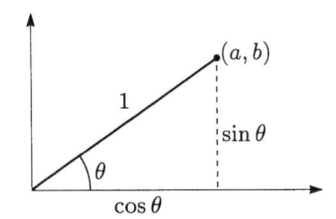

$$(a, b) = (\cos\theta, \sin\theta).$$

Figure 5.1

A similar argument shows that (c, d) can be written as

$$(c, d) = (\cos\phi, \sin\phi),$$

for some ϕ.

Now the image of $(1, 0)$ is (a, b), and the image of $(0, 1)$ is (c, d). So far we have used only the fact that the images of $(1, 0)$ and $(0, 1)$ are of unit length. We may also use the fact that they are at right angles. This implies that

$$\phi = \theta \pm \pi/2.$$

But if $\phi = \theta + \pi/2$, then

$$(\cos\phi, \sin\phi) = (-\sin\theta, \cos\theta),$$

whereas if $\phi = \theta - \pi/2$, then

$$(\cos\phi, \sin\phi) = (\sin\theta, -\cos\theta).$$

Thus the matrix $\mathbf{A}$ has one of the two forms

$$\begin{bmatrix} \cos\theta & -\sin\theta \\ \sin\theta & \cos\theta \end{bmatrix} \quad \text{or} \quad \begin{bmatrix} \cos\theta & \sin\theta \\ \sin\theta & -\cos\theta \end{bmatrix}$$

for some angle θ. The first form is that of a rotation about the origin through the angle θ, as illustrated in Figure 5.2(a). This form has determinant $+1$. The second form is that of a reflection in a line which contains the origin and is inclined at an angle $\theta/2$ to the x-axis (measured in the anticlockwise sense), as illustrated in Figure 5.2(b). This form has determinant -1.

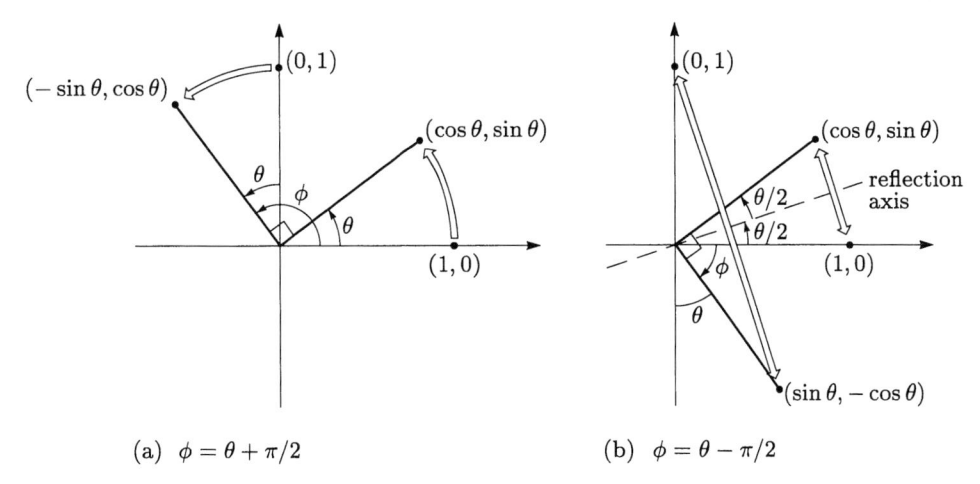

(a) $\phi = \theta + \pi/2$ (b) $\phi = \theta - \pi/2$

Figure 5.2

To see why the reflection axis is inclined at an angle $\theta/2$ to the x-axis, look at Figure 5.2(b). Since the reflection of $(1,0)$ to $(\cos\theta, \sin\theta)$ is equivalent to moving $(1,0)$ through an angle θ, there must be an angle $\theta/2$ between the line from $(0,0)$ to $(1,0)$ and the reflection axis, and another angle $\theta/2$ between the reflection axis and the line from $(0,0)$ to the image of $(1,0)$.

This difference between determinant $+1$ and determinant -1 is a fundamental one.

Definition 5.1 Direct and indirect isometries

An isometry whose linear part has transformation matrix with determinant $+1$ is a **direct isometry**; one whose linear part has transformation matrix with determinant -1 is an **indirect isometry**.

If you model a plane isometry by physically manipulating a sheet of paper, then the direct isometries do not involve turning the paper over, while the indirect ones do.

The anticlockwise rotation about the origin through the angle θ, achieved by applying the matrix

$$\begin{bmatrix} \cos\theta & -\sin\theta \\ \sin\theta & \cos\theta \end{bmatrix}, \quad \text{will be denoted by } r[\theta],$$

while the reflection in the line through the origin inclined at an anticlockwise angle θ to the x-axis, corresponding to the matrix

$$\begin{bmatrix} \cos 2\theta & \sin 2\theta \\ \sin 2\theta & -\cos 2\theta \end{bmatrix}, \quad \text{will be denoted by } q[\theta].$$

It is important to note that the θ in the symbol $q[\theta]$ refers to the angle of the reflection axis, *not* the angle by which $(1,0)$ is moved. We could have defined the notation the other way, but because there are two possibilities, somebody is going to be unhappy whatever we do! If you are the unhappy one, please bear with us!

Exercise 5.1

Consider the square S whose corners are at $(1,1)$, $(1,-1)$, $(-1,-1)$ and $(-1,1)$. Using the above notation, write down all the isometries that map S to itself.

Now, we have discovered that the linear part of every isometry can be written in the form $r[\theta]$ or $q[\theta]$ for some θ. Let us next consider the geometric effect of the isometries $r[\theta]$ and $q[\theta]$, in terms of those points in the plane that are left fixed by each of them.

The operation $r[\theta]$ (for any angle θ that is not a multiple of 2π) clearly leaves just the origin fixed, and no other point. On the other hand, the operation $e = r[0]$ fixes *every point in the plane*. The operation $q[\theta]$ (for *any* angle θ, *including $\theta = 0$*) fixes every point on a line through the origin, namely the line of reflection, but no other point.

As is usual, we denote the identity operation by e. Thus
$$r[0] = t[\mathbf{0}] = e,$$
but notice that $q[0] \neq e$.

Now what if we compose each isometry of the form $r[\theta]$ or $q[\theta]$ with a translation, $t[\mathbf{p}]$? What are the possibilities?

First, we consider the *direct* isometries.

For example, consider the isometry f obtained by composing $r[\pi/2]$ with a translation by 1 unit in the x-direction:

$$f = t[\mathbf{p}]\, r[\pi/2], \qquad \text{where } \mathbf{p} = (1, 0).$$

We have composed an isometry which leaves just one point fixed with an isometry which fixes no points at all. What is the result? The answer is that f is still an anticlockwise rotation through a right angle, but the centre of rotation (in other words, the fixed point) is no longer the origin. So where is it? It is not hard to set up a pair of simultaneous equations to find out. If $\mathbf{x} = (x, y)$ is the fixed point, then $r[\pi/2]$, which has transformation matrix

$$\begin{bmatrix} 0 & -1 \\ 1 & 0 \end{bmatrix},$$

maps (x, y) to $(-y, x)$. Then, translating by $(1, 0)$ moves $(-y, x)$ to $(-y+1, x)$, and this point must be the same as the original point (x, y). So we have

$$-y + 1 = x$$
$$x = y$$

and so

$$x = y = \tfrac{1}{2}.$$

You can easily check that $\left(\tfrac{1}{2}, \tfrac{1}{2}\right)$ is indeed the fixed point of f, by applying the transformation matrix followed by the translation by $(1, 0)$ to it. The result of doing this is illustrated in Figure 5.3(a). If we reduce the picture in Figure 5.3(a) to a simple triangle, as shown in Figure 5.3(b), we can see that **the side of the triangle opposite the right angle represents the translation back to the fixed point, and must therefore have the direction and length of the vector $(1, 0)$.**

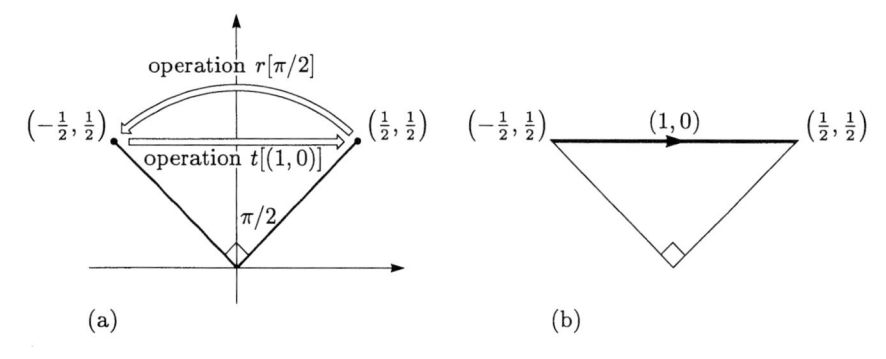

Figure 5.3

If, in this example, we had chosen the origin of the coordinate system to be at the point which (in the present system) has the coordinates $\left(\frac{1}{2}, \frac{1}{2}\right)$, then the isometry f would simply have been a rotation through a right angle; that is, it would have had the same form as does the isometry $r[\theta]$ in the current coordinate system. *Geometrically*, these two isometries are thus of the same form. It is convenient to have a notation for such a rotation. We write $r[\mathbf{c}, \theta]$ for the anticlockwise rotation through θ about the point $\mathbf{c}$. Thus, the above isometry is denoted by $r\left[\left(\frac{1}{2}, \frac{1}{2}\right), \pi/2\right]$.

The above discussion is easy to generalize. For any non-zero rotation $r[\theta]$ and any translation $t[\mathbf{p}]$, the composite isometry $r[\mathbf{c}, \theta] = t[\mathbf{p}] \, r[\theta]$ has exactly one fixed point $\mathbf{c}$, which can be found by writing the isometry algebraically and solving the corresponding equations. This composite isometry is simply a rotation through θ about the point $\mathbf{c}$.

If we now note that, by Theorem 3.2 and Definition 5.1, all direct isometries of the plane can be written in the form $t[\mathbf{p}] \, r[\theta]$, for some vector $\mathbf{p}$ and angle θ, then it follows that there are just three types of direct isometry of the plane:

(a) the identity isometry e, which fixes every point; $\qquad e = t[\mathbf{0}] \, r[0]$.

(b) non-zero translations $t[\mathbf{p}]$ (where $\mathbf{p} \neq \mathbf{0}$), which do not fix any point in the plane; $\qquad t[\mathbf{p}] = t[\mathbf{p}] \, r[0]$.

(c) non-zero rotations $r[\mathbf{c}, \theta]$ (where $\theta \neq 0$) about some point $\mathbf{c}$ in the plane, which fix just the single point $\mathbf{c}$. $\qquad r[\mathbf{c}, \theta] = t[\mathbf{p}] \, r[\theta]$, for some $\mathbf{p}$.

Next, we consider the *indirect* isometries.

Let us start with the reflection $q[0]$. In this case, the entire x-axis consists of fixed points, while any point not on the x-axis has the sign of its (non-zero) y-coordinate reversed, and is therefore not a fixed point.

What if we now compose $q[0]$ with a translation? In fact, there is more than one possibility, as the following exercise shows.

Exercise 5.2

Let $\mathbf{p} = (1, 0), \mathbf{r} = (0, -1)$ and $\mathbf{s} = (1, -2)$. What are the fixed points of:

(a) $f = t[\mathbf{p}] \, q[0]$?

(b) $g = t[\mathbf{r}] \, q[0]$?

(c) $h = t[\mathbf{s}] \, q[0]$?

We can deduce, from Exercise 5.2, that if we compose a reflection $q[\theta]$, for *any* angle θ, with a translation *at right angles* to the reflection axis, we get a reflection in an axis parallel to the first (and whose fixed points consist of this new axis), whereas if we compose a reflection $q[\theta]$ with a translation having a component *parallel* to the reflection axis, the result is an isometry having no fixed points at all. Furthermore, in the latter case, we can also deduce that the composite isometry consists of a reflection followed by a (non-zero) translation parallel to the line of reflection. This new type of isometry has a special name.

Definition 5.2 Glide reflection

A **glide reflection** is an indirect isometry consisting of a reflection followed by a non-zero translation parallel to the line of reflection.

As with rotations, we now need to set up some notation for geometrically described reflections and glide reflections. Suppose we have a reflection in the line which passes through the point whose vector is **c** and which is inclined at an anticlockwise angle θ ($0 \leq \theta < \pi$) to the x-axis. We denote this reflection by $q[\mathbf{c}, \theta]$. Suppose now that we follow $q[\mathbf{c}, \theta]$ by a non-zero translation $t[\mathbf{g}]$ parallel to the line of reflection. Thus we have a glide reflection, whose line of reflection is still the line through **c** inclined at angle θ. This glide reflection will be denoted by $q[\mathbf{g}, \mathbf{c}, \theta]$, and the line of reflection through **c** is referred to as its *glide reflection axis*.

Note that the expression $q[\mathbf{c}, \theta]$ is not unique. For example, the line inclined at $\pi/4$ to the x-axis and passing through $(-1, 0)$ also passes through $(0, 1)$, so $q[(-1, 0), \pi/4]$ and $q[(0, 1), \pi/4]$ are different notations for the same reflection. In fact, since there are infinitely many points on any line, there are infinitely many ways of denoting any reflection using this notation.

Exercise 5.3

Use the above notation to describe:

(a) reflection in the line $x + y = 2$;

(b) reflection in the line $x + y = 2$, followed by translation by one unit parallel to this line in the direction of increasing y.

The letters, $\mathbf{g}$, $\mathbf{c}$, θ are written in this order because we always compose operations from right to left. The operation $q[\mathbf{g}, \mathbf{c}, \theta]$ is constructed by taking a line through the origin inclined at an angle θ, moving it so that it passes through $\mathbf{c}$, then reflecting about the line, then translating by $\mathbf{g}$.

If we now note that, by Theorem 3.2 and Definition 5.1, all indirect isometries of the plane can be written in the form $t[\mathbf{p}] \, q[\theta]$, for some vector **p** and angle θ, then it follows that there are just two types of indirect isometry of the plane:

(a) reflections $q[\mathbf{c}, \theta]$, which fix the points on the reflection axis;

(b) glide reflections $q[\mathbf{g}, \mathbf{c}, \theta]$ (where $\mathbf{g} \neq \mathbf{0}$), which do not fix any point in the plane.

Glide reflections share with translations the property that no *point* is fixed. Yet it is clear that their geometric properties are quite different from those of translations. A good way to confirm this is to look at their effects on *lines* (straight lines, that is) in the plane.

Although a glide reflection fixes no *point* in the plane, it does keep just one *line* fixed as a whole, namely the glide reflection axis itself. The points on this axis are moved along it (i.e. the axis is moved along itself), but the line itself is fixed. Also, a glide reflection has the effect of mapping lines parallel to the glide reflection axis from one side of the axis to the other.

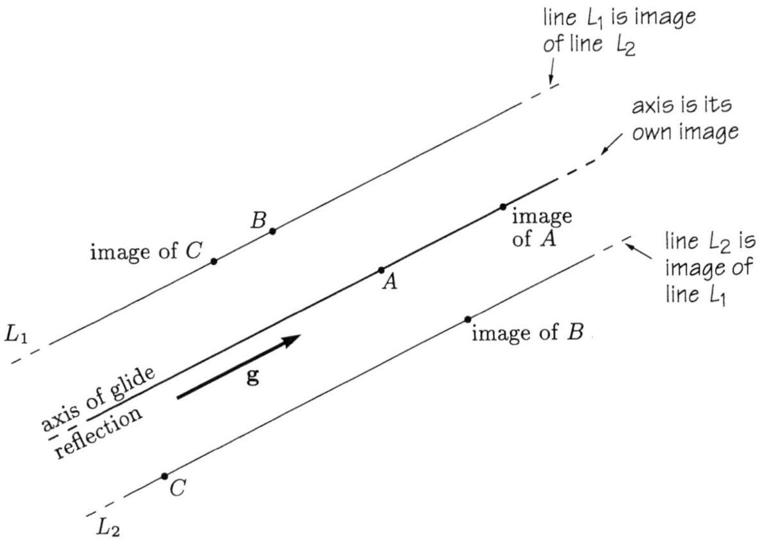

Figure 5.4 The effect of a glide reflection on lines in the plane.

A translation, on the other hand, keeps *every line parallel to the direction of translation* fixed as a whole.

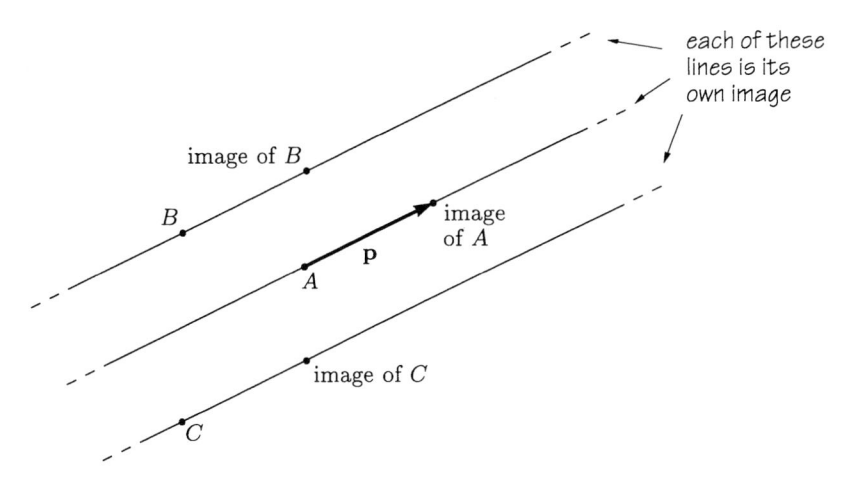

Figure 5.5 The effect of a translation on lines in the plane.

As we have looked at the effects on lines in order to make a geometric distinction between translations and glide reflections, we should now consider the effects on lines of the other three types of isometry.

The *identity isometry* clearly maps every line to itself. Furthermore, every point on every line maps exactly to itself.

Most non-zero *rotations* clearly do not map any lines onto themselves. However, there is a subtlety: a rotation through exactly π about a point C maps *every line through C* to itself as a whole (though the only fixed *point* is C itself). Thus, if we define the geometric properties of an isometry in terms of fixed lines as well as fixed points, a *rotation through exactly π is a new geometric type in its own right*.

The case of *reflections* is slightly less obvious. Let q be a reflection in a line L. We have seen that every point on L remains fixed; so, of course, L as a whole remains fixed. But this is not all; it is also true that *every line perpendicular to L* maps to itself as a whole. (For example, the reflection $q[0]$ reverses the sign of the y-coordinate of any point in the plane, while keeping the x-coordinate constant; so any line of the form $x = k$ maps, as a whole, to itself.)

To summarize this subsection, we have found *four* geometric types of direct isometry and *two* geometric types of indirect isometry. We collect the results together in the following theorem.

> **Theorem 5.1 Geometric characterization of isometries**
>
> The six geometric types of isometry in the plane are:
>
> (a) The *identity*, denoted by e.
>
>   This isometry is direct, and fixes every point and every line in the plane.
>
> (b) *Non-zero translation by* $\mathbf{p}$ (where $\mathbf{p} \neq \mathbf{0}$), denoted by $t[\mathbf{p}]$.   — If $\mathbf{p} = \mathbf{0}$, we get type (a).
>
>   An isometry of this type is direct, fixes no points, but fixes as a whole every line parallel to the direction of translation.
>
> (c) *Rotation through an angle θ other than 0 or π about a point C*, denoted by $r[\mathbf{c}, \theta]$, where $\mathbf{c}$ is the vector representing the point C and where $\theta \in [0, 2\pi[$ is measured in an anticlockwise direction.   — If $\theta = 0$, we get type (a). If $\theta = \pi$, we get type (d). Sometimes we may prefer to consider $\theta \in\,]-\pi, \pi]$.
>
>   An isometry of this type is direct, fixes the point C and no other, and fixes no lines.
>
> (d) *Rotation through π about a point C*, denoted by $r[\mathbf{c}, \pi]$, where $\mathbf{c}$ is a vector representing the point C.
>
>   An isometry of this type is direct, fixes the point C and no other, and fixes as a whole every line through C.
>
> (e) *Reflection in a line L*, denoted by $q[\mathbf{c}, \theta]$, where $\mathbf{c}$ represents any point on L and where $\theta \in [0, \pi[$ represents the angle (measured in an anticlockwise direction) that L makes with the x-axis.
>
>   An isometry of this type is indirect, fixes every point on L (and therefore fixes L as a whole), and fixes as a whole every line perpendicular to L.
>
> (f) *Glide reflection*, consisting of reflection in a line L followed by non-zero translation parallel to L, and denoted by $q[\mathbf{g}, \mathbf{c}, \theta]$, where $\mathbf{c}$ represents any point on the line L, $\theta \in [0, \pi[$ represents the angle (measured in an anticlockwise direction) that L makes with the x-axis and $\mathbf{g}$ represents the parallel translation.
>
>   An isometry of this type is indirect, fixes no points, and fixes the line L as a whole.

Having classified isometries geometrically, we are now in a position to determine the geometric effect of composing isometries. This forms part of the discussion of the next subsection.

5.2 The Isometry Toolkit

The Isometry Toolkit is simply a set of equations and diagrams for manipulating isometries. It is printed on a card in the *Geometry Envelope*. The remainder of this section is a brief description of how these equations are derived, together with some exercises on the use of the Toolkit.

You will need to refer to the Isometry Toolkit card throughout this subsection.

Equation 1 of the Toolkit is simply Rule 1 of Section 3.

Equations 2–5 can be derived from Rule 2 of Section 3, by writing down the matrices for $r[\theta]$, etc., and multiplying them together. But they are also quite easy to justify geometrically.

You would also need to use various trigonometric identities to derive these equations directly from the matrices.

Equation 2 simply says that a rotation through θ, followed by a rotation through ϕ, constitutes a rotation through $\theta + \phi$.

Equation 3 can be derived by first noting that the composite of two reflections through the origin, having determinant $(-1) \times (-1) = 1$, must be a rotation about the origin. Now, the x-axis is reflected by $q[\phi]$ to the line inclined at an angle 2ϕ to the x-axis. This line makes an angle $2\phi - \theta$ with the reflection axis of $q[\theta]$, so it is reflected to the line making an angle $-(2\phi - \theta)$ with this axis. This is the line making an angle $\theta - (2\phi - \theta) = 2(\theta - \phi)$ with the x-axis, and so the required rotation is $r[2(\theta - \phi)]$.

Recall that $\det \mathbf{AB} = \det \mathbf{A} \det \mathbf{B}$. Also recall, from Theorem 3.1, that the composite of isometries is an isometry. Recall further, from Subsection 5.1, that the transformation matrix of an isometry can only have determinant $+1$ (for a linear part that is a rotation) or -1 (for a linear part that is a reflection).

Equations 4 and 5 can be derived similarly, by noting that the results must be reflections, and tracing where the x-axis goes.

Equation 6 is simply a re-expression of Rule 3 of Section 3. Equations 6a and 6b write this out in more detail. Whether you find the concise form of Equation 6 or the more explicit form of Equations 6a and 6b more congenial is up to you.

Now we shall consider the cases of rotations about points other than the origin and reflections in lines that do not pass through the origin.

We wish to bring $r[\mathbf{c}, \theta]$ to standard form. The first thing to notice is that, if we translate the point $\mathbf{c}$ to the origin by the translation $t[-\mathbf{c}]$, rotate by θ using $r[\theta]$, then translate the origin back to $\mathbf{c}$ by $t[\mathbf{c}]$, the result is $r[\mathbf{c}, \theta]$:

$$r[\mathbf{c}, \theta] = t[\mathbf{c}]\, r[\theta]\, t[-\mathbf{c}].$$

This is Equation 7 of the Toolkit.

Next, using Equations 6a and 1 of the Toolkit, we get

$$\begin{aligned} t[\mathbf{c}]\, (r[\theta]\, t[-\mathbf{c}]) &= t[\mathbf{c}]\, (t[-r[\theta](\mathbf{c})]\, r[\theta]) \quad \text{(by Equation 6a)} \\ &= (t[\mathbf{c}]\, t[-r[\theta](\mathbf{c})])\, r[\theta] \\ &= t[\mathbf{d}]\, r[\theta], \quad \text{where } \mathbf{d} = \mathbf{c} - r[\theta](\mathbf{c}) \quad \text{(by Equation 1).} \end{aligned}$$

This is Equation 8 of the Toolkit.

Writing g for $t[\mathbf{c}]$ and x for $r[\theta]$, we observe that $r[\mathbf{c}, \theta]$ becomes gxg^{-1}. You may well recognize this type of relationship from your previous studies: $r[\mathbf{c}, \theta]$ is conjugate *to $r[\theta]$ in the group of all plane isometries; and, in any group, conjugate elements make similar contributions to the group structure. In particular, conjugacy in a geometric group implies geometric similarity.*

That is to say, to find the vector $\mathbf{d}$ such that $r[\mathbf{c}, \theta]$ takes the standard form $t[\mathbf{d}]\, r[\theta]$, we rotate $\mathbf{c}$ by θ and subtract the result from $\mathbf{c}$. Figure 5.6 shows a diagram of the situation.

$$\mathbf{d} = \mathbf{c} - r[\theta](\mathbf{c})$$

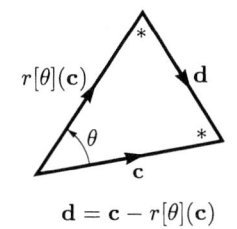

Figure 5.6

Figure 5.6 shows an isosceles triangle, the sides $\mathbf{c}$ and $r[\theta](\mathbf{c})$ being of equal length. Thus the angles marked $*$ are equal.

In the particular case when $\theta = \pi$, illustrated in Figure 5.7, we have $r[\theta](\mathbf{c}) = -\mathbf{c}$, and so

$$r[\mathbf{c}, \pi] = t[2\mathbf{c}]\, r[\pi].$$

This is Equation 9 of the Toolkit.

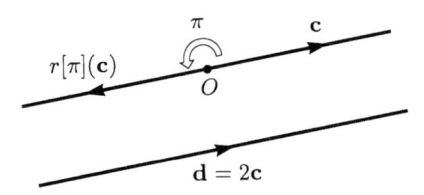

Figure 5.7

How do we reverse the process? That is, given an expression in the standard form $t[\mathbf{d}]\, r[\theta]$, how do we find the rotation centre $\mathbf{c}$ — the vector $\mathbf{c}$ such that $t[\mathbf{d}]\, r[\theta] = r[\mathbf{c}, \theta]$?

One way is to set up a pair of simultaneous equations, as we did in Subsection 5.1. Alternatively, we can use a diagram such as Figure 5.6: draw the vector $\mathbf{d}$, then erect an isosceles triangle with base $\mathbf{d}$ and apex angle θ, and finally take the apex as the origin.

Exercise 5.4

Find the centre of rotation of $t[(0, 2)]\, r[\pi/2]$.

Next, consider reflections and glide reflections. Arguing exactly as in the case of rotations, but this time making use of Equation 6b rather than Equation 6a, we have

$$\begin{aligned} q[\mathbf{c}, \theta] &= t[\mathbf{c}]\, q[\theta]\, t[-\mathbf{c}] \\ &= t[\mathbf{d}]\, q[\theta], \qquad \text{where } \mathbf{d} = \mathbf{c} - q[\theta](\mathbf{c}). \end{aligned}$$

These are Equations 10 and 11 of the Toolkit.

Again, we can illustrate the situation in a diagram, such as Figure 5.8

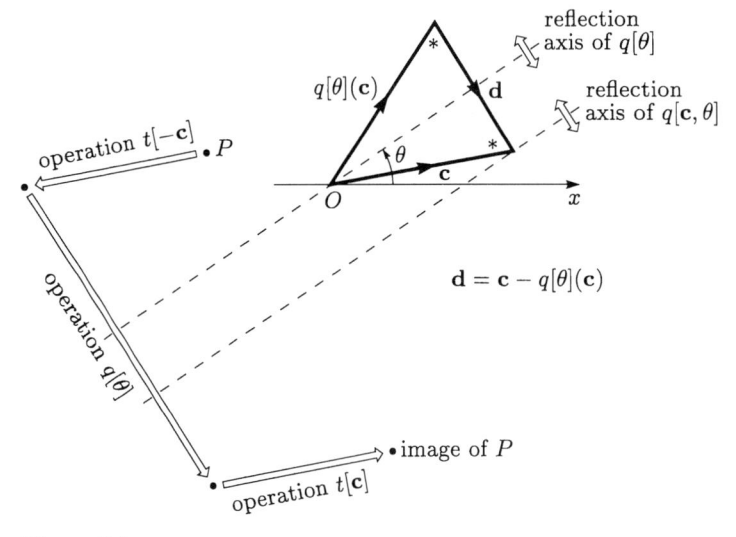

Figure 5.8

In the particular case when **c** is chosen to be perpendicular to the reflection axis (that is, **c** represents the point on the reflection axis that is closest to the origin), we can deduce that

$$q[\mathbf{c}, \theta] = t[2\mathbf{c}]\, q[\theta],$$

as Figure 5.9 illustrates. This is Equation 12 of the Toolkit.

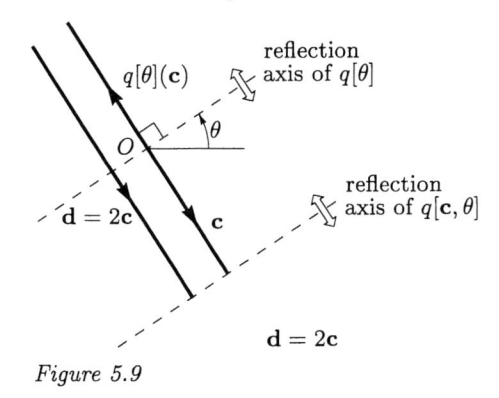

Figure 5.9

Once again, given an expression for a reflection in the standard form $t[\mathbf{d}]\, q[\theta]$, it is possible to obtain the vector **c** that defines the reflection axis by setting up a pair of simultaneous equations. Alternatively, we can use a diagram such as Figure 5.8.

Now consider the glide reflection $q[\mathbf{g}, \mathbf{c}, \theta]$:

$$\begin{aligned}
q[\mathbf{g}, \mathbf{c}, \theta] &= t[\mathbf{g}]\, q[\mathbf{c}, \theta] &&\text{(by definition)}\\
&= t[\mathbf{g}]\, t[\mathbf{c}]\, q[\theta]\, t[-\mathbf{c}] &&\text{(by Equation 10)}\\
&= t[\mathbf{g} + \mathbf{c}]\, q[\theta]\, t[-\mathbf{c}] &&\text{(by Equation 1)}\\
&= t[\mathbf{g} + \mathbf{c}]\, t[-q[\theta](\mathbf{c})]\, q[\theta] &&\text{(by Equation 6b)}\\
&= t[\mathbf{g} + \mathbf{c} - q[\theta](\mathbf{c})]\, q[\theta] &&\text{(by Equation 1)}\\
&= t[\mathbf{d}]\, q[\theta], &&\text{where } \mathbf{d} = \mathbf{g} + \mathbf{c} - q[\theta](\mathbf{c}).
\end{aligned}$$

Thus we have Equations 13 and 14 of the Toolkit.

Furthermore, when **c** is perpendicular to the reflection axis, so that $q[\theta](\mathbf{c}) = -\mathbf{c}$, we can deduce that

$$q[\mathbf{g}, \mathbf{c}, \theta] = t[\mathbf{g} + 2\mathbf{c}]\, q[\theta].$$

This is Equation 15 of the Toolkit.

The following expressions for the inverses of our plane isometries, which form Equations 16–21 of the Toolkit, are trivial consequences of the definitions of translations, rotations, reflections and glide reflections:

$$\begin{aligned}
(t[\mathbf{p}])^{-1} &= t[-\mathbf{p}]\\
(r[\theta])^{-1} &= r[-\theta] = r[2\pi - \theta]\\
(q[\theta])^{-1} &= q[\theta]\\
(r[\mathbf{c}, \theta])^{-1} &= r[\mathbf{c}, -\theta] = r[\mathbf{c}, 2\pi - \theta]\\
(q[\mathbf{c}, \theta])^{-1} &= q[\mathbf{c}, \theta]\\
(q[\mathbf{g}, \mathbf{c}, \theta])^{-1} &= q[-\mathbf{g}, \mathbf{c}, \theta]
\end{aligned}$$

Exercise 5.5

(a) Express $t[(2, -1)]\, r[\pi]$ as a rotation about some point.

(b) Express the rotation $r[(0, 1), \pi/2]$ in standard form.

Exercise 5.6

(a) Express $r[\pi]\, t[\mathbf{p}]\, r[\pi]$ in standard form.

(b) Express $r[\pi/2]\, t[\mathbf{p}]\, r[-\pi/2]$ in standard form.

Exercise 5.7

At the beginning of the section, we asked 'What is the result of performing first an anticlockwise rotation through $\pi/3$ about the point $(\sqrt{3}, 1)$, and then a reflection in the line $y = x + 3$?' Answer this question by giving an expression in standard form. Write down a simple geometric description of this expression.

Hint Use Equation 12 of the Toolkit to obtain an expression for the reflection in $y = x + 3$ in standard form.

Exercise 5.8

(a) Express the glide reflection $q[(1, -1), (1, 1), 3\pi/4]$ in standard form.

(b) Describe geometrically the glide reflection whose expression in standard form is $t[(1,1)] \, q[\pi/2]$, i.e. describe the glide reflection as reflection in a certain line followed by translation along this line by a certain amount.

Hint Use Equation 15 of the Toolkit in both parts.

Exercise 5.9

Consider the following squares:

S, whose corners are at $(0,0), (1,0), (1,1), (0,1)$;

T, whose corners are at $(0,0), (1,0), (1,-1), (0,-1)$.

Express in standard form:

(a) the rotation through π which maps S to T;

(b) the reflection which maps S to T;

(c) the glide reflection which maps S to T.

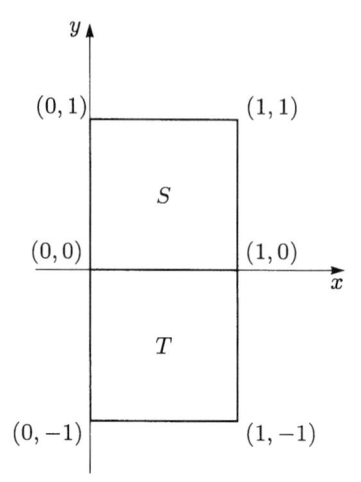

Figure 5.10

Conversion of isometries to explicit form

Throughout the rest of this course, our normal notation for isometries will be the one used above, and we shall perform manipulations using Equations 1–21 of the Isometry Toolkit. However, you may prefer to use the **explicit form** of an isometry, in terms of x and y, which you met in Section 3, given by

$$(x, y) \longmapsto (ax + cy + u, bx + dy + v)$$

or equivalently

$$\begin{bmatrix} x \\ y \end{bmatrix} \longmapsto \begin{bmatrix} a & c \\ b & d \end{bmatrix} \begin{bmatrix} x \\ y \end{bmatrix} + \begin{bmatrix} u \\ v \end{bmatrix}.$$

Example 5.1

Suppose you are asked to find the result of performing, first, the glide reflection whose axis is the line $x = 1$ and the glide part of which is translation by 2 units vertically upwards, then the rotation through π about the point $(0, 1)$.

In the notation developed above, the glide reflection is $q[(0,2),(1,0),\pi/2]$ and the rotation is $r[(0,1),\pi]$. Thus the composite is

$$r[(0,1),\pi]\ q[(0,2),(1,0),\pi/2]$$
$$= t[2(0,1)]\ r[\pi]\ t[(0,2) + 2(1,0)]\ q[\pi/2] \quad \text{(by Equations 9 and 15)}$$
$$= t[(0,2)]\ r[\pi]\ t[(2,2)]\ q[\pi/2]$$
$$= t[(0,2)]\ t[(-2,-2)]\ r[\pi]\ q[\pi/2] \quad \text{(by Equation 6a)}$$
$$= t[(-2,0)]\ q[\pi/2 + \pi/2] \quad \text{(by Equations 1 and 4)}$$
$$= t[(-2,0)]\ q[\pi]$$
$$= t[(-2,0)]\ q[0].$$

Alternatively, you could use Equation 22 of the Isometry Toolkit to convert $t[(2,2)]\ q[\pi/2]$ to

$$(x,y) \mapsto (x\cos\pi + y\sin\pi + 2, x\sin\pi - y\cos\pi + 2) = (-x + 2, y + 2),$$

or equivalently (using Equation 22a) to

$$\begin{bmatrix} x \\ y \end{bmatrix} \mapsto \begin{bmatrix} -1 & 0 \\ 0 & 1 \end{bmatrix} \begin{bmatrix} x \\ y \end{bmatrix} + \begin{bmatrix} 2 \\ 2 \end{bmatrix},$$

and you could use Equation 23 of the Isometry Toolkit to convert $t[(0,2)]\ r[\pi]$ to

$$(x,y) \mapsto (x\cos\pi - y\sin\pi, x\sin\pi + y\cos\pi + 2) = (-x, -y + 2),$$

or equivalently (using Equation 23a) to

$$\begin{bmatrix} x \\ y \end{bmatrix} \mapsto \begin{bmatrix} -1 & 0 \\ 0 & -1 \end{bmatrix} \begin{bmatrix} x \\ y \end{bmatrix} + \begin{bmatrix} 0 \\ 2 \end{bmatrix}.$$

Then the composite is

$$(x,y) \mapsto \bigl(-(-x+2), -(y+2) + 2\bigr) = (x - 2, -y),$$

or equivalently

$$\begin{bmatrix} x \\ y \end{bmatrix} \mapsto \begin{bmatrix} -1 & 0 \\ 0 & -1 \end{bmatrix} \left(\begin{bmatrix} -1 & 0 \\ 0 & 1 \end{bmatrix} \begin{bmatrix} x \\ y \end{bmatrix} + \begin{bmatrix} 2 \\ 2 \end{bmatrix} \right) + \begin{bmatrix} 0 \\ 2 \end{bmatrix}$$
$$= \begin{bmatrix} 1 & 0 \\ 0 & -1 \end{bmatrix} \begin{bmatrix} x \\ y \end{bmatrix} + \begin{bmatrix} -1 & 0 \\ 0 & -1 \end{bmatrix} \begin{bmatrix} 2 \\ 2 \end{bmatrix} + \begin{bmatrix} 0 \\ 2 \end{bmatrix}$$
$$= \begin{bmatrix} 1 & 0 \\ 0 & -1 \end{bmatrix} \begin{bmatrix} x \\ y \end{bmatrix} + \begin{bmatrix} -2 \\ 0 \end{bmatrix}. \qquad \blacklozenge$$

APPENDIX: PROOF OF THE FUNDAMENTAL THEOREM OF AFFINE GEOMETRY

The material in this appendix is optional.

> **Fundamental theorem of affine geometry**
>
> Let P, Q, R be any three non-collinear points in $\mathbb{R}^2$, and let U, V, W be any three other such points. Then there is exactly one affine transformation that maps P to U, Q to V and R to W.

Proof

Let the coordinates of the points P, Q and R be given by, respectively,

$$\mathbf{p} = (p_1, p_2), \quad \mathbf{q} = (q_1, q_2), \quad \mathbf{r} = (r_1, r_2).$$

Let $q_1 - p_1 = a$, $q_2 - p_2 = b$, and let $r_1 - p_1 = c$, $r_2 - p_2 = d$, so that

$$\mathbf{q} - \mathbf{p} = (a, b) \quad \text{and} \quad \mathbf{r} - \mathbf{p} = (c, d).$$

Since P, Q and R are non-collinear, the vectors $\mathbf{q} - \mathbf{p}$ and $\mathbf{r} - \mathbf{p}$ are linearly independent, and therefore the matrix

$$\mathbf{A} = \begin{bmatrix} a & c \\ b & d \end{bmatrix}$$

is invertible.

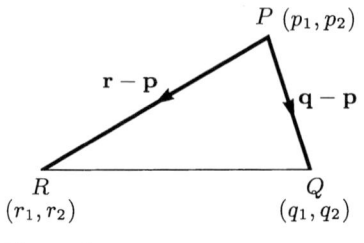

Figure A.1

Now let $A = (0,0)$, $B = (1,0)$, $C = (0,1)$, and consider the effect on these three points of the affine transformation

$$f = t[\mathbf{p}] \, \lambda[\mathbf{A}].$$

You should have no difficulty in checking that

$$f(A) = P, \quad f(B) = Q, \quad f(C) = R.$$

Moreover, f is the *only* affine transformation with this property, since a transformation with a different translation part would clearly map A to some point other than P, whereas a transformation $t[\mathbf{p}] \, \lambda[\mathbf{B}]$ with $\mathbf{B} \neq \mathbf{A}$ would map B to a point other than Q (if the first column of $\mathbf{B}$ differed from the first column of $\mathbf{A}$) or would map C to a point other than R (if the second column of $\mathbf{B}$ differed from the second column of $\mathbf{A}$).

Next, by exactly the same argument as above, there is exactly one affine transformation g such that

$$g(A) = U, \quad g(B) = V, \quad g(C) = W.$$

Therefore the composite affine transformation gf^{-1} maps P to U, Q to V and R to W. Moreover, since f and g are unique, so is gf^{-1}. ∎

Corollary

Any triangle can be mapped exactly onto any other triangle, and any parallelogram can be mapped exactly onto any other parallelogram, by a suitable affine transformation.

Proof

The result concerning triangles is an obvious direct consequence of the theorem.

To prove the result concerning parallelograms, we begin by letting X and Y be any two parallelograms. Any three vertices P, Q, R of X form a triangle, as do any three vertices P', Q', R' of Y. Suppose we have chosen them in such a way that PR is a diagonal of X and $P'R'$ is a diagonal of Y. Let S and S' be the fourth vertices of X and Y respectively.

Let h be the affine transformation which sends P, Q and R to P', Q' and R' respectively. Since affine transformations send parallel lines to parallel lines, it follows that $h(S)$ lies on a line through P' parallel to $Q'R'$, and also on a line through R' parallel to $P'Q'$. Hence $h(S) = S'$, and so h maps X to Y, as required. ∎

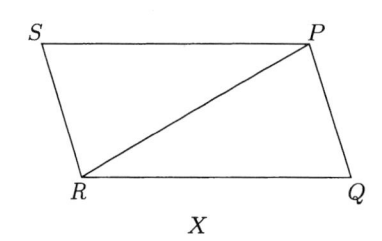

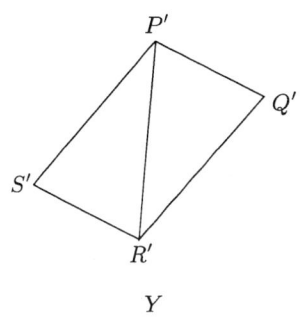

Figure A.2

SOLUTIONS TO THE EXERCISES

Solution 1.1
Shapes (a), (b) and (e) are topological discs. Shape (c) is not because it has a hole. Shapes (d) and (f) are not because each can be separated into two pieces by deleting a single point.

Solution 1.2
Sets (b) and (d) are topological discs. Set (a) is not for two reasons: first, it is a line and so can be separated into two by deleting a single point; second, it is infinite. Set (c) is also not a topological disc, since it is a (circular) line and so can be separated into two by deleting just two points.

Set (c), however, *encloses* a topological disc, namely the set of points whose distance from the fixed point is at most 2.5 cm.

Solution 1.3
Pictures (a) to (g) inclusive and (l) represent tilings. The others do not: (h) because the tiles overlap; (i) because the cloister-shaped tiles are not topological discs; (j) because it covers only an elliptical portion of the plane, not the whole plane; and (k) because there are gaps between the tiles, so again the whole plane is not covered.

Our solution assumes that the pictures in Figure 1.1 can all be extended over the plane in a manner suggested by the pictures.

Solution 1.4
As a tiling is defined to cover the whole plane, a boundary point must divide one tile from another and not from nothing! So it must belong to at least two tiles.

Solution 1.5
(a) The vertices on the boundary of T_5 are V_2, V_3, V_6, V_7; the edges are E_4, E_7, E_8, E_{11}.
(b) The edge E_8 lies on the boundaries of tiles T_5 and T_6.
(c) The vertex V_3 lies on the boundaries of tiles T_2, T_3, T_5 and T_6.

Solution 1.6
(a) Tiles T_2, T_4, T_6 and T_8 are adjacent to T_5.
(b) No, because although they share a common *vertex*, they do not share a common *edge*.
(c) No, because although they share a common *vertex*, they do not bound a common *tile*.

Solution 1.7
(a) 4.
(b) 4.
(In fact, the answer is 4 for any tiling at all!)

Solution 1.8
Vertices V_2 and V_3, and tiles T_2 and T_5.

Solution 1.9
The vertices are all of degree 3. Some tiles are of degree 4 and some are of degree 8.

Solution 2.1
Tilings (a), (c), (e), (g) and (l) are polygonal.

Solution 2.2

One possible example is given in Figure S.1.

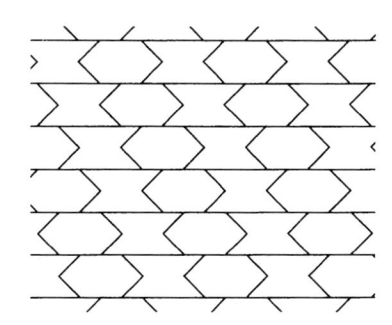

Figure S.1

This one repeats a regular pattern; but of course there are plenty that do not even do that!

Solution 2.3

Tilings (b) and (c) are edge-to-edge.

Solution 2.4

A regular hexagon can be divided into six equilateral triangles by drawing in its three diagonals, as shown in Figure S.2. Therefore, starting with $\mathcal{R}_6$, we can choose a subset of the hexagonal tiles and divide them in this way. Any such choice of a subset of the hexagonal tiles to divide in this way produces an edge-to-edge tiling; thus, infinitely many different edge-to-edge tilings by regular polygons can be constructed.

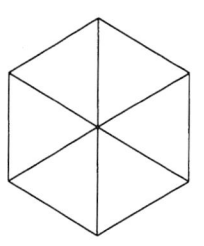

Figure S.2

One such possibility is given in Figure S.3.

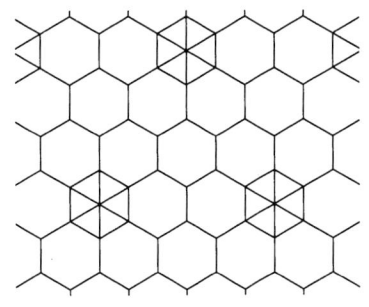

Figure S.3

This one repeats a regular pattern; but, again, there are plenty that do not even do that!

Solution 2.5

The first tiling has vertex types $(3,3,3,3,6)$ and $(3,3,6,6)$; the second has vertex types $(3,3,3,4,4)$ and $(3,3,4,3,4)$.

Solution 2.6

$(3,3,3,3,3,3)$, $(4,4,4,4)$ and $(6,6,6)$.

Solution 2.7

$(3,3,3,3,3,3)$, $(4,4,4,4)$ and $(3,6,3,6)$.

Solution 3.1

$$\begin{aligned}
(g \circ f)(\mathbf{x}) &= g(f(\mathbf{x})) \\
&= g(\mathbf{Ax} + \mathbf{p}) \\
&= \mathbf{B}(\mathbf{Ax} + \mathbf{p}) + \mathbf{q} \\
&= \mathbf{BAx} + (\mathbf{Bp} + \mathbf{q}).
\end{aligned}$$

Thus $g \circ f$ is an affine transformation given by

$$g \circ f = t[\mathbf{r}] \circ \lambda[\mathbf{C}],$$

where $\mathbf{r} = \mathbf{Bp} + \mathbf{q}$ and $\mathbf{C} = \mathbf{BA}$.

Solution 3.2

The transformations are:

$$f^{-1}: \mathbf{x} \mapsto \begin{bmatrix} 0 & 1 \\ -1 & 0 \end{bmatrix} \mathbf{x} + \begin{bmatrix} 1 \\ 1 \end{bmatrix}$$

$$f^2 : \mathbf{x} \mapsto \begin{bmatrix} -1 & 0 \\ 0 & -1 \end{bmatrix} \mathbf{x} + \begin{bmatrix} 2 \\ 0 \end{bmatrix}$$

$$f^3 : \mathbf{x} \mapsto \begin{bmatrix} 0 & 1 \\ -1 & 0 \end{bmatrix} \mathbf{x} + \begin{bmatrix} 1 \\ 1 \end{bmatrix}$$

$$f^4 : \mathbf{x} \mapsto \begin{bmatrix} 1 & 0 \\ 0 & 1 \end{bmatrix} \mathbf{x} + \begin{bmatrix} 0 \\ 0 \end{bmatrix}$$

That is to say:

$$\begin{aligned} f^{-1}(x,y) &= (y+1, -x+1) \\ f^2(x,y) &= (-x+2, -y) \\ f^3(x,y) &= (y+1, -x+1) \\ f^4(x,y) &= (x,y) \end{aligned}$$

Solution 3.3

A drawing is given in Figure S.4. The shape is a parallelogram.

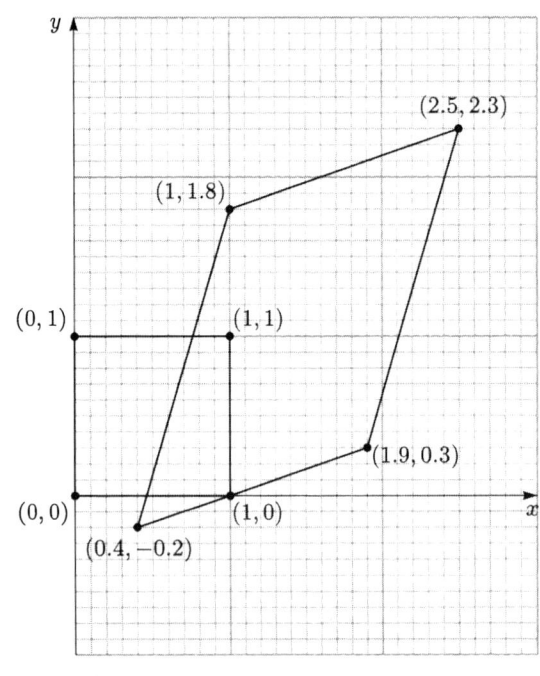

Figure S.4

Solution 3.4

(a) This function is not an isometry. For example, let $\mathbf{p} = (1,1)$, $\mathbf{q} = (1,-1)$. Then $\mathbf{p} - \mathbf{q} = (0,2)$, and so

$$\|\mathbf{p} - \mathbf{q}\| = 2.$$

However, $f(\mathbf{p}) = f(\mathbf{q}) = (1,1)$, so

$$\|f(\mathbf{p}) - f(\mathbf{q})\| = 0 \neq \|\mathbf{p} - \mathbf{q}\|.$$

(b) This function is an isometry (as you might expect, since it is a translation). To prove this, take any two points $\mathbf{p} = (a,b)$, $\mathbf{q} = (c,d)$. Then
$$\mathbf{p} - \mathbf{q} = (a-c, b-d).$$
However,
$$f(\mathbf{p}) = (a+1, b+1), \quad f(\mathbf{q}) = (c+1, d+1),$$
so that
$$\begin{aligned} f(\mathbf{p}) - f(\mathbf{q}) &= (a+1-c-1, b+1-d-1) \\ &= (a-c, b-d) \\ &= \mathbf{p} - \mathbf{q}, \end{aligned}$$
and so
$$\|f(\mathbf{p}) - f(\mathbf{q})\| = \|\mathbf{p} - \mathbf{q}\|.$$

(c) This function is not an isometry. For example, let $\mathbf{p} = (1,1)$, $\mathbf{q} = (0,0)$. Then $\mathbf{p} - \mathbf{q} = (1,1)$, and so
$$\|\mathbf{p} - \mathbf{q}\| = \sqrt{2}.$$
However, $f(\mathbf{p}) = (2,0)$ and $f(\mathbf{q}) = (0,0)$, so
$$\|f(\mathbf{p}) - f(\mathbf{q})\| = 2 \neq \|\mathbf{p} - \mathbf{q}\|.$$

(d) This function is an isometry. To prove this, take any two points $\mathbf{p} = (a,b)$, $\mathbf{q} = (c,d)$. Then
$$\mathbf{p} - \mathbf{q} = (a-c, b-d).$$
Also
$$f(\mathbf{p}) - f(\mathbf{q}) = (b-d, a-c),$$
so clearly
$$\|f(\mathbf{p}) - f(\mathbf{q})\| = \|\mathbf{p} - \mathbf{q}\|.$$

It is a reflection in the line $y = x$.

(e) This function is an isometry. Once again, take any two points $\mathbf{p} = (a,b)$, $\mathbf{q} = (c,d)$. Then
$$\mathbf{p} - \mathbf{q} = (a-c, b-d),$$
and so
$$\|\mathbf{p} - \mathbf{q}\|^2 = (a-c)^2 + (b-d)^2.$$
However,
$$f(\mathbf{p}) = \left(\tfrac{1}{2}a - \tfrac{1}{2}b\sqrt{3}, \tfrac{1}{2}a\sqrt{3} + \tfrac{1}{2}b\right), \quad f(\mathbf{q}) = \left(\tfrac{1}{2}c - \tfrac{1}{2}d\sqrt{3}, \tfrac{1}{2}c\sqrt{3} + \tfrac{1}{2}d\right),$$
so that
$$f(\mathbf{p}) - f(\mathbf{q}) = \left(\tfrac{1}{2}(a-c) - \tfrac{1}{2}(b-d)\sqrt{3}, \tfrac{1}{2}(a-c)\sqrt{3} + \tfrac{1}{2}(b-d)\right),$$
and so
$$\|f(\mathbf{p}) - f(\mathbf{q})\|^2 = \left(\tfrac{1}{2}(a-c) - \tfrac{1}{2}(b-d)\sqrt{3}\right)^2 + \left(\tfrac{1}{2}(a-c)\sqrt{3} + \tfrac{1}{2}(b-d)\right)^2,$$
which simplifies to
$$(a-c)^2 + (b-d)^2.$$
Therefore taking positive square roots (since distances must be positive), we get
$$\|f(\mathbf{p}) - f(\mathbf{q})\| = \|\mathbf{p} - \mathbf{q}\|.$$

It is an anticlockwise rotation through the angle $\pi/3$ about the origin.

Solution 3.5

All the functions except (a) are affine transformations.

In (b), the linear part is the identity transformation (with transformation matrix the identity matrix, $\mathbf{I}$) and the translation part is translation by $(1,1)$.

In each of (c)–(e), the translation part is the identity translation (i.e. translation by the zero vector, $t[\mathbf{0}]$), while the linear parts have the following transformation matrices:

(c) $\begin{bmatrix} 1 & 1 \\ 1 & -1 \end{bmatrix}$

(d) $\begin{bmatrix} 0 & 1 \\ 1 & 0 \end{bmatrix}$

(e) $\begin{bmatrix} \frac{1}{2} & -\frac{1}{2}\sqrt{3} \\ \frac{1}{2}\sqrt{3} & \frac{1}{2} \end{bmatrix}$

Solution 3.6

For any two vectors $\mathbf{q}$, $\mathbf{r}$, and for $f = t[\mathbf{p}]$, we have

$$\|f(\mathbf{q}) - f(\mathbf{r})\| = \|(\mathbf{p}+\mathbf{q}) - (\mathbf{p}+\mathbf{r})\| = \|\mathbf{q}-\mathbf{r}\|,$$

and so f is an isometry.

Solution 3.7

Let $\mathbf{p} = (a,b)$, $\mathbf{q} = (c,d)$, $\mathbf{r} = (e,f)$.

(a) $\mathbf{p} \cdot \mathbf{q} = ac + bd$
$= ca + db$
$= \mathbf{q} \cdot \mathbf{p}$

(b) $(\mathbf{p}+\mathbf{q}) \cdot \mathbf{r} = (a+c, b+d) \cdot (e,f)$
$= (a+c)e + (b+d)f$
$= (ae+bf) + (ce+df)$
$= \mathbf{p} \cdot \mathbf{r} + \mathbf{q} \cdot \mathbf{r}$

(c) $\mathbf{p} \cdot (\mathbf{q}+\mathbf{r}) = (a,b) \cdot (c+e, d+f)$
$= a(c+e) + b(d+f)$
$= (ac+bd) + (ae+bf)$
$= \mathbf{p} \cdot \mathbf{q} + \mathbf{p} \cdot \mathbf{r}$

Solution 3.8

For any isometry f and any two vectors $\mathbf{p}$, $\mathbf{q}$, from Equation 3.5 we have

$$\bigl(f(\mathbf{p}) - f(\mathbf{q})\bigr) \cdot \bigl(f(\mathbf{p}) - f(\mathbf{q})\bigr) = (\mathbf{p}-\mathbf{q}) \cdot (\mathbf{p}-\mathbf{q}).$$

Using the result of Exercise 3.7(c), we get

$$\bigl(f(\mathbf{p}) - f(\mathbf{q})\bigr) \cdot f(\mathbf{p}) - \bigl(f(\mathbf{p}) - f(\mathbf{q})\bigr) \cdot f(\mathbf{q}) = (\mathbf{p}-\mathbf{q}) \cdot \mathbf{p} - (\mathbf{p}-\mathbf{q}) \cdot \mathbf{q},$$

and, using the result of Exercise 3.7(b), we get

$$f(\mathbf{p}) \cdot f(\mathbf{p}) - f(\mathbf{q}) \cdot f(\mathbf{p}) - f(\mathbf{p}) \cdot f(\mathbf{q}) + f(\mathbf{q}) \cdot f(\mathbf{q}) = \mathbf{p} \cdot \mathbf{p} - \mathbf{q} \cdot \mathbf{p} - \mathbf{p} \cdot \mathbf{q} + \mathbf{q} \cdot \mathbf{q};$$

and so, using the result of Exercise 3.7(a), we have

$$f(\mathbf{p}) \cdot f(\mathbf{p}) + f(\mathbf{q}) \cdot f(\mathbf{q}) - 2 f(\mathbf{p}) \cdot f(\mathbf{q}) = \mathbf{p} \cdot \mathbf{p} + \mathbf{q} \cdot \mathbf{q} - 2\mathbf{p} \cdot \mathbf{q}. \quad \text{(S.1)}$$

Now, we are assuming that the origin is preserved, i.e. that $f(\mathbf{0}) = \mathbf{0}$.
Therefore,
$$\begin{aligned} f(\mathbf{p}) \cdot f(\mathbf{p}) &= \|f(\mathbf{p})\|^2 \\ &= \|f(\mathbf{p}) - f(\mathbf{0})\|^2 \\ &= \|\mathbf{p} - \mathbf{0}\|^2 \\ &= \|\mathbf{p}\|^2 \\ &= \mathbf{p} \cdot \mathbf{p}, \end{aligned}$$
and similarly,
$$f(\mathbf{q}) \cdot f(\mathbf{q}) = \mathbf{q} \cdot \mathbf{q}.$$
Hence, using these results together with Equation S.1, we get
$$f(\mathbf{p}) \cdot f(\mathbf{q}) = \mathbf{p} \cdot \mathbf{q}, \tag{S.2}$$
as required.

Further, if θ is the angle between $\mathbf{p}$ and $\mathbf{q}$ and if ϕ is the angle between $f(\mathbf{p})$ and $f(\mathbf{q})$, then, using Equation 3.4, we have
$$\begin{aligned} \cos \phi &= f(\mathbf{p}) \cdot f(\mathbf{q}) / \sqrt{(f(\mathbf{p}) \cdot f(\mathbf{p}))(f(\mathbf{q}) \cdot f(\mathbf{q}))} \\ &= \mathbf{p} \cdot \mathbf{q} / \sqrt{(\mathbf{p} \cdot \mathbf{p})(\mathbf{q} \cdot \mathbf{q})} \quad \text{(by Equation S.2)} \\ &= \cos \theta. \end{aligned}$$

Therefore, for angles $\theta, \phi \in [0, \pi[$, we must have $\phi = \theta$, and so magnitudes of angles are preserved.

> Remember that angles between vectors are specified as angles in $[0, \pi[$.

Solution 3.9

The transformation matrices are:

(a) $\begin{bmatrix} 1 & -1 \\ 1 & 1 \end{bmatrix}$, which is not orthogonal, as $a^2 + b^2 = c^2 + d^2 = 2$, not 1;

(b) $\begin{bmatrix} 0 & 1 \\ -1 & 0 \end{bmatrix}$, which is orthogonal;

(c) $\begin{bmatrix} 1/\sqrt{2} & 1/\sqrt{2} \\ 1/\sqrt{2} & -1/\sqrt{2} \end{bmatrix}$, which is orthogonal.

Solution 3.10

(a) Let $\mathbf{A}$ and $\mathbf{B}$ be orthogonal matrices. Then $\lambda[\mathbf{A}]$ and $\lambda[\mathbf{B}]$ are isometries, by Theorem 3.2 (with $t[\mathbf{p}] = t[\mathbf{0}]$). Therefore $\lambda[\mathbf{A}] \circ \lambda[\mathbf{B}]$ is an isometry, by Theorem 3.1. But this is equal to $\lambda[\mathbf{AB}]$, by the composition rule for affine transformations (Rule 4), and so $\mathbf{AB}$ is orthogonal, by Theorem 3.2.

(b) Let $\mathbf{A}$ be an orthogonal matrix. Then $\lambda[\mathbf{A}]$ is an isometry, by Theorem 3.2. Therefore $(\lambda[\mathbf{A}])^{-1}$ is an isometry, by Theorem 3.1. But this is equal to $\lambda[\mathbf{A}^{-1}]$, and so $\mathbf{A}^{-1}$ is orthogonal, by Theorem 3.2.

Solution 4.1

Tilings (a), (b), (c) and (d) are monohedral.

Solution 4.2

The two orientations differ by a rotation through $\pi/2$.

Solution 4.3

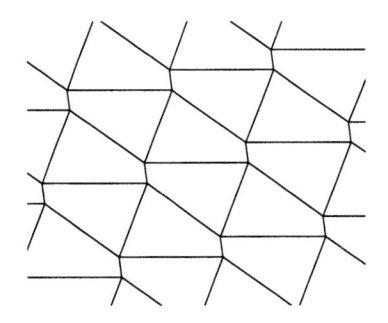

Figure S.5

In this tiling, every tile can be obtained from any of its adjacent neighbours by rotating by a half turn about the common edge. Notice that, by rotating the original template about any of its sides, the new shape together with the original form a parallelohexagon, translated copies of which tile the plane.

Solution 4.4

(a) [3, 3, 4, 3, 4]

(c) [3, 3, 3, 3, 3, 3]

(d) [3, 3, 3, 3, 3, 3]

(e) [3, 3, 3, 3, 3, 3]

(f) [3, 3, 3, 4, 4]

Solution 4.5

(3, 3, 3, 3, 3, 3) is tile-uniform, with tile type [6, 6, 6].
(4, 4, 4, 4) is tile-uniform, with tile type [4, 4, 4, 4].
(6, 6, 6) is tile-uniform, with tile type [3, 3, 3, 3, 3, 3].
That is to say, the tile type of $\mathcal{R}_3$ is essentially the vertex type of $\mathcal{R}_6$, and vice versa, while the tile and vertex types of $\mathcal{R}_4$ are essentially the same.

The other Archimedean tilings are not tile-uniform.

Notice that a prerequisite for tile-uniformity is that each tile must have the same number of vertices.

Solution 4.6

The thin lines represent the tiling $\mathcal{R}_6$.

Solution 4.7

The dual is just $\mathcal{R}_4$ again. Thus $\mathcal{R}_4$ is *self-dual*.

Solution 4.8

The three regular tilings are Laves tilings and Archimedean tilings.

Solution 4.9

[4, 4, 4, 4], [6, 6, 6], [4, 6, 12] and [4, 8, 8].

Solution 4.10

[3, 3, 3, 3, 3, 3]: degree 6.
[3, 3, 3, 3, 6], [3, 3, 3, 4, 4], [3, 3, 4, 3, 4]: degree 5.
[4, 4, 4, 4], [3, 4, 6, 4], [3, 6, 3, 6]: degree 4.
[6, 6, 6], [3, 12, 12], [4, 6, 12], [4, 8, 8]: degree 3.

Solution 5.1

There are eight isometries:

$$r[0]\ (= e,\text{ the identity isometry}),\quad r[\pi/2],\quad r[\pi],\quad r[3\pi/2];$$
$$q[0],\quad q[\pi/4],\quad q[\pi/2],\quad q[3\pi/4].$$

In words, these isometries are rotations through multiples of $\pi/2$ and reflections in the x- and y-axes and in the diagonals of the square.

Note that, for rotations in the plane, we normally restrict ourselves to angles in the principal interval $[0, 2\pi[$. Similarly, for reflections, we normally restrict ourselves to angles in the principal interval $[0, \pi[$. Sometimes, however, it will be more convenient to use angles in the intervals $]-\pi, \pi]$ and $]-\pi/2, \pi/2]$ respectively.

Solution 5.2

(a) Regardless of what happens to the y-coordinates, every point has its x-coordinate increased by 1. Thus there are no fixed points. If you try to find the solution by using simultaneous equations, you get

$$x + 1 = x, \quad -y = y,$$

and the first equation is impossible to satisfy.

Notice that the composite isometry in (a) is a reflection in the line $y = 0$ followed by a translation parallel to this line.

(b) The equations this time are

$$x = x, \quad -y - 1 = y.$$

The solution set is the line $y = -\frac{1}{2}$; that is to say, the fixed points are those on the line through $\left(0, -\frac{1}{2}\right)$ parallel to the x-axis.

Notice that the composite isometry in (b) is simply a reflection in the line $y = -\frac{1}{2}$.

(c) The equations this time are

$$x + 1 = x, \quad -y - 2 = y.$$

As in part (a), the first equation is impossible to satisfy, so there are no fixed points.

Notice that the composite isometry in (c) is a reflection in the line $y = -1$ followed by a translation parallel to this line.

Solution 5.3

(a) The line $x + y = 2$ is inclined at an anticlockwise angle of $3\pi/4$ to the x-axis. Thus, for any vector $\mathbf{c}$ on this line, $q[\mathbf{c}, 3\pi/4]$ is a suitable notation to describe this reflection. For example, $q[(2,0), 3\pi/4], q[(1,1), 3\pi/4]$ and $q[(0,2), 3\pi/4]$ are all correct.

(b) From Figures S.6 and S.7 we can deduce that $\mathbf{g} = (-\sqrt{2}/2, \sqrt{2}/2)$. So, for example,

$$q[(-\sqrt{2}/2, \sqrt{2}/2), (2, 0), 3\pi/4],$$
$$q[(-\sqrt{2}/2, \sqrt{2}/2), (1, 1), 3\pi/4],$$
$$q[(-\sqrt{2}/2, \sqrt{2}/2), (0, 2), 3\pi/4]$$

are all correct.

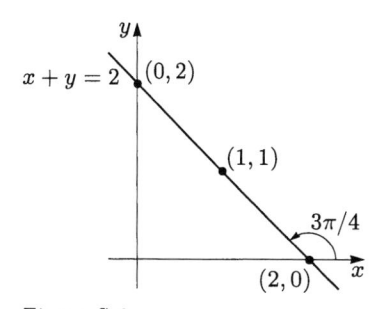

Figure S.6

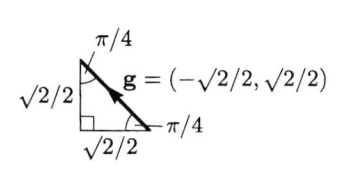

Figure S.7

Solution 5.4

The equations for the centre of rotation are

$$-y = x, \quad x + 2 = y,$$

whose solution is $x = -1, y = 1$. Thus the rotation centre is $(-1, 1)$.

Alternatively, the result can be obtained from Figure S.8.

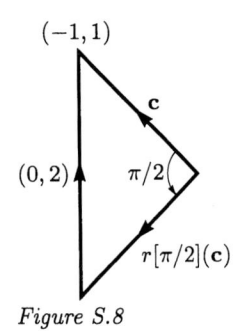

Figure S.8

(Notice that the other possible isosceles triangle that could have been constructed with $\mathbf{d} = (0, 2)$ as its base, shown in Figure S.9, is incorrect, since if we position and orient the vectors $\mathbf{c}$ and $r[\pi/2](\mathbf{c})$ to match the orientations in Figure 5.6, then the angle between $\mathbf{c}$ and $r[\pi/2](\mathbf{c})$ is measured in a *clockwise* direction.)

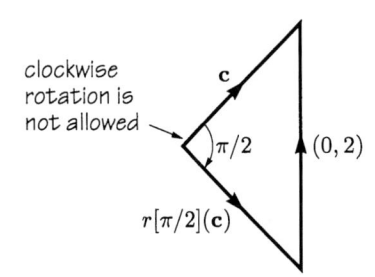

Figure S.9

Solution 5.5

(a) Using Equation 9 of the Toolkit, we seek $r[\mathbf{c}, \pi]$, where $2\mathbf{c} = (2, -1)$. Thus, the rotation is $r\left[\left(1, -\tfrac{1}{2}\right), \pi\right]$, rotation through π about the point $\left(1, -\tfrac{1}{2}\right)$.

(b) Using Equation 8 of the Toolkit with $\theta = \pi/2$ and $\mathbf{c} = (0, 1)$, we have

$$\mathbf{d} = (0, 1) - r[\pi/2](0, 1) = (1, 1),$$

so the required expression is $t[(1, 1)]\, r[\pi/2]$.

Solution 5.6

(a) $\begin{aligned}[t] r[\pi]\, t[\mathbf{p}]\, r[\pi] &= (r[\pi]\, t[\mathbf{p}])\, r[\pi] \\ &= (t[-\mathbf{p}]\, r[\pi])\, r[\pi] \quad \text{(by Equation 6a)} \\ &= t[-\mathbf{p}]\, (r[\pi]\, r[\pi]) \\ &= t[-\mathbf{p}] \quad \text{(by Equation 2).} \end{aligned}$

(b) $\begin{aligned}[t] r[\pi/2]\, t[\mathbf{p}]\, r[-\pi/2] &= (r[\pi/2]\, t[\mathbf{p}])\, r[-\pi/2] \\ &= (t[\mathbf{q}]\, r[\pi/2])\, r[-\pi/2] \quad \text{where } \mathbf{q} = r[\pi/2](\mathbf{p}) \quad \text{(by Equation 6a)} \\ &= t[\mathbf{q}]\, (r[\pi/2]\, r[-\pi/2]) \\ &= t[\mathbf{q}] \quad \text{(by Equation 2).} \end{aligned}$

Solution 5.7

An anticlockwise rotation through $\pi/3$ about $(\sqrt{3}, 1)$ has the expression $r[(\sqrt{3},1), \pi/3]$. Now,

$$r[\pi/3](\sqrt{3},1) = (0,2),$$

as Figure S.10 illustrates. So (using Equation 8 of the Toolkit), we find that this rotation has the standard form $t[\mathbf{d}]\, r[\pi/3]$, where

$$\mathbf{d} = (\sqrt{3},1) - (0,2) = (\sqrt{3}, -1).$$

Next, the line $y = x + 3$ passes through the point $(-3/2, 3/2)$, and the vector $(-3/2, 3/2)$ is perpendicular to the reflection axis (see Figure S.11). Thus we can use Equation 12 of the Toolkit to express the reflection in the given line as

$$t[(-3,3)]\, q[\pi/4].$$

Thus, an expression for the result of performing first the given rotation and then the given reflection is

$$t[(-3,3)]\, q[\pi/4]\, t[(\sqrt{3}, -1)]\, r[\pi/3].$$

Now, $q[\pi/4]$, which represents reflection in the line $y = x$, simply exchanges x- and y-coordinates. Thus, $q[\pi/4](\sqrt{3}, -1) = (-1, \sqrt{3})$. Therefore, using Equation 6b of the Toolkit, we can rewrite our expression as

$$t[(-3,3)]\, t[(-1, \sqrt{3})]\, q[\pi/4]\, r[\pi/3].$$

Using Equations 1 and 5 of the Toolkit, we find that this gives the standard form

$$t[(-4, 3+\sqrt{3})]\, q[\pi/4 - \pi/6] = t[(-4, 3+\sqrt{3})]\, q[\pi/12].$$

Geometrically, this is a reflection in the line passing through the origin and inclined at an angle $\pi/12$ to the x-axis, followed by the translation by $\mathbf{p} = (-4, 3+\sqrt{3})$.

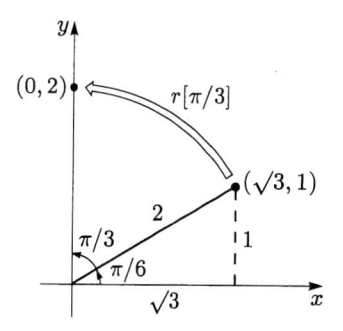

Figure S.10

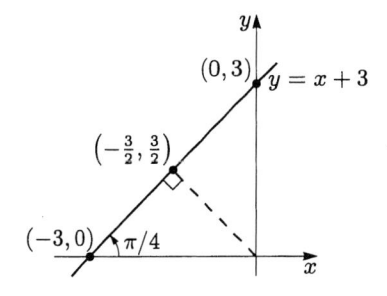

Figure S.11

Solution 5.8

(a) In this case, $\mathbf{c} = (1,1)$ is perpendicular to the reflection axis, so we can use Equation 15 of the Toolkit. We thus obtain the standard form

$$t[(3,1)]\, q[3\pi/4].$$

(b) First note that the expression represents a glide reflection in an axis parallel to the y-axis. So in order to use Equation 15 of the Toolkit we must express $(1,1)$ as $\mathbf{g} + 2\mathbf{c}$, where $\mathbf{c}$ is perpendicular to the y-axis and where $\mathbf{g}$ is parallel to the y-axis. Clearly, $\mathbf{g} = (0,1)$ and $\mathbf{c} = \left(\frac{1}{2}, 0\right)$ is the correct choice, so that

$$t[(1,1)]\, q[\pi/2] = q\left[(0,1), \left(\tfrac{1}{2}, 0\right), \pi/2\right].$$

This represents reflection in the line parallel to the y-axis through $\left(\frac{1}{2}, 0\right)$ followed by translation along this axis by one unit in the direction of increasing y.

Solution 5.9

(a) We can see from Figure 5.10 that the rotation centre must be the midpoint of the common side of S and T, namely $\left(\frac{1}{2}, 0\right)$. Thus the required rotation is $r\left[\left(\frac{1}{2}, 0\right), \pi\right]$, whose standard form is $t[(1,0)]\, r[\pi]$ (by Equation 9 of the Toolkit).

(b) This reflection is reflection in the x-axis, namely $q[0]$.

(c) The glide reflection which maps S to T is reflection in the vertical line bisecting the squares, followed by translation by one unit in the direction of decreasing y. That is to say, it is $q\left[(0,-1), \left(\frac{1}{2}, 0\right), \pi/2\right]$, whose standard form is $t[(1,-1)]\, q[\pi/2]$ (by Equation 15 of the Toolkit).

Other rotations are possible for mapping S to T, but there is only one through π.

OBJECTIVES

After you have studied this unit, you should be able to:

(a) explain what is meant by a *tiling*, recognize drawings of tilings and interpret them correctly;

(b) explain what is meant by the *net* of a tiling and by the *parts* of a tiling, and determine the *adjacency* and *incidence* properties of a given tiling;

(c) define *polygonal* tilings and *edge-to-edge* tilings;

(d) recognize the *Archimedean* tilings, understand their description in terms of *vertex types*, and classify the Archimedean tilings into the three *regular* and the eight *semi-regular* tilings;

(e) construct *non-Archimedean* tilings using regular polygons;

(f) define and recognize *monohedral tilings*, and use various methods to construct them;

(g) recognize the eleven *Laves* tilings, understand their description in terms of *tile types*, and understand the *duality* between these and the Archimedean tilings;

(h) distinguish those functions from the plane to itself that are *affine transformations*, and write an affine transformation as a composite of a *linear part* and a *translation part*;

(i) recognize those affine transformations that are *isometries*, as having linear parts that are described by *orthogonal* matrices;

(j) classify plane isometries into six types, find the type of a given isometry, and say, for each type, which points it fixes, and which lines it fixes both pointwise and as a whole;

(k) use the Isometry Toolkit to manipulate isometries.

INDEX

adjacent edges 14
adjacent tiles 14
adjacent vertices 14
affine congruency lemma 35
affine transformation 24
algebraic characterization of
    isometries 30
alternate angles 37
angle between vectors 28
Archimedean tiling 19, 20
Archimedean tiling theorem 21
composition rule for affine
    transformations 25
congruence 19, 32
convex polygon 16
coordinate system 23
corner of polygon 16
degree of tile 15
degree of vertex 15
direct isometry 43
distance 27
divides 19
dot product 28
dual tilings 40
edge-to-edge tiling 17
edge of tiling 12
equality of tile types 39
equality of vertex types 20
explicit form of isometry 52
fundamental theorem of affine
    geometry 33, 54

geometric characterization of
    isometries 48
glide reflection 45
incident parts of tiling 15
indirect isometry 43
inverse rule for affine
    transformations 25
isometry 26
    direct 43
    explicit form of 52
    indirect 43
    standard form of 42
Isometry Toolkit 42, 49
Laves tiling 40
length of vector 27, 28
linear part of affine transformation 24
linear transformation 23
matrix of transformation 23
monohedral tiling 32
monohedral tiling theorem 36
net of tiling 11
non-collinearity 33
orthogonal matrix 29
parallelogram 35
parallelohexagon 35
parts of tiling 13
plane isometry 42
polygon 16
polygonal tiling 16
regular polygon 18
regular tiling 18, 20

regular tiling theorem 19
semi-regular tiling 19, 20
side of polygon 16
standard basis vectors 23
standard form of isometry 42
template 32
tile 10
tile type 38
tile-uniform tiling 39
tiling 10
    Archimedean 19, 20
    edge-to-edge 17
    Laves 40
    monohedral 32
    polygonal 16
    regular 18, 20
    semi-regular 19, 20
    vertex-uniform 20
topological disc 8
transformation
    affine 24
    linear 23
translation part of affine
    transformation 24
unit square 26
unit vector 27
vertex of tiling 12
vertex type 20
vertex-uniform tiling 20